SEQUENCING AND SCHEDULING:
An Introduction to the Mathematics of the Job-Shop

ELLIS HORWOOD SERIES IN
MATHEMATICS AND ITS APPLICATIONS

Series Editor: Professor G. M. BELL, Chelsea College, University of London

The works in this series will survey recent research, and introduce new areas and up-to-date mathematical methods. Undergraduate texts on established topics will stimulate student interest by including present-day applications, and the series can also include selected volumes of lecture notes on important topics which need quick and early publication.

In all three ways it is hoped to render a valuable service to those who learn, teach, develop and use mathematics.

MATHEMATICAL THEORY OF WAVE MOTION
G. R. BALDOCK and T. BRIDGEMAN, University of Liverpool.
MATHEMATICAL MODELS IN SOCIAL MANAGEMENT AND LIFE SCIENCES
D. N. BURGHES and A. D. WOOD, Cranfield Institute of Technology.
MODERN INTRODUCTION TO CLASSICAL MECHANICS AND CONTROL
D. N. BURGHES, Cranfield Institute of Technology and A. DOWNS, Sheffield University.
CONTROL AND OPTIMAL CONTROL
D. N. BURGHES, Cranfield Institute of Technology and A. GRAHAM, The Open University, Milton Keynes.
TEXTBOOK OF DYNAMICS
F. CHORLTON, University of Aston, Birmingham.
VECTOR AND TENSOR METHODS
F. CHORLTON, University of Aston, Birmingham.
TECHNIQUES IN OPERATIONAL RESEARCH
VOLUME 1: QUEUEING SYSTEMS
VOLUME 2: MODELS, SEARCH, RANDOMIZATION
B. CONNOLLY, Chelsea College, University of London
MATHEMATICS FOR THE BIOSCIENCES
G. EASON, C. W. COLES, G. GETTINBY, University of Strathclyde.
HANDBOOK OF HYPERGEOMETRIC INTEGRALS: Theory, Applications, Tables, Computer Programs
H. EXTON, The Polytechnic, Preston.
MULTIPLE HYPERGEOMETRIC FUNCTIONS
H. EXTON, The Polytechnic, Preston
COMPUTATIONAL GEOMETRY FOR DESIGN AND MANUFACTURE
I. D. FAUX and M. J. PRATT, Cranfield Institute of Technology.
APPLIED LINEAR ALGEBRA
R. J. GOULT, Cranfield Institute of Technology.
MATRIX THEORY AND APPLICATIONS FOR ENGINEERS AND MATHEMATICIANS
A. GRAHAM, The Open University, Milton Keynes.
APPLIED FUNCTIONAL ANALYSIS
D. H. GRIFFEL, University of Bristol.
GENERALISED FUNCTIONS: Theory, Applications
R. F. HOSKINS, Cranfield Institute of Technology.
MECHANICS OF CONTINUOUS MEDIA
S. C. HUNTER, University of Sheffield.
GAME THEORY: Mathematical Models of Conflict
A. J. JONES, Royal Holloway College, University of London.
USING COMPUTERS
B. L. MEEK and S. FAIRTHORNE, Queen Elizabeth College, University of London.
SPECTRAL THEORY OF ORDINARY DIFFERENTIAL OPERATORS
E. MULLER-PFEIFFER, Technical High School, Ergurt.
SIMULATION CONCEPTS IN MATHEMATICAL MODELLING
F. OLIVEIRA-PINTO, Chelsea College, University of London.
ENVIRONMENTAL AERODYNAMICS
R. S. SCORER, Imperial College of Science and Technology, University of London.
APPLIED STATISTICAL TECHNIQUES
K. D. C. STOODLEY, T. LEWIS and C. L. S. STAINTON, University of Bradford.
LIQUIDS AND THEIR PROPERTIES: A Molecular and Macroscopic Treatise with Applications
H. N. V. TEMPERLEY, University College of Swansea, University of Wales and D. H. TREVENA, University of Wales, Aberystwyth.
GRAPH THEORY AND APPLICATIONS
H. N. V. TEMPERLEY, University College of Swansea.

SEQUENCING AND SCHEDULING: An Introduction to the Mathematics of the Job-Shop

SIMON FRENCH, B.A., M.A., D.Phil.
Department of Decision Theory
University of Manchester

ELLIS HORWOOD LIMITED
Publishers · Chichester

Halsted Press: a division of
JOHN WILEY & SONS
New York · Chichester · Brisbane · Toronto

First published in 1982 by

ELLIS HORWOOD LIMITED
Market Cross House, Cooper Street, Chichester, West Sussex, PO19 1EB, England

The publisher's colophon is reproduced from James Gillison's drawing of the ancient Market Cross, Chichester.

Distributors:

Australia, New Zealand, South-east Asia:
Jacaranda-Wiley Ltd., Jacaranda Press,
JOHN WILEY & SONS INC.,
G.P.O. Box 859, Brisbane, Queensland 40001, Australia

Canada:
JOHN WILEY & SONS CANADA LIMITED
22 Worcester Road, Rexdale, Ontario, Canada.

Europe, Africa:
JOHN WILEY & SONS LIMITED
Baffins Lane, Chichester, West Sussex, England.

North and South America and the rest of the world:
Halsted Press: a division of
JOHN WILEY & SONS
605 Third Avenue, New York, N.Y. 10016, U.S.A.

© **1982 S. French/Ellis Horwood Ltd.**

British Library Cataloguing in Publication Data
French, Simon
Sequencing and scheduling. — (Ellis Horwood series in mathematics and its applications)
1. Scheduling (Management)
I. Title
658.5'1   TS157.A2

Library of Congress Card No. 81-3414 AACR2

ISBN 0-85312-299-7 (Ellis Horwood Ltd., Publishers — Library Edn.)
ISBN 0-85312-364-0 (Ellis Horwood Ltd., Publishers — Student Edn.)
ISBN 0-470-27229-5 (Halsted Press)

Typeset in Great Britain by Preface, Salisbury.
Printed in the USA by The Maple-Vail Book Manufacturing Group, New York.

Table of Contents

To Judy

Preface

I first realised the need for a new textbook on the mathematics of scheduling when I was asked to develop and teach a course on this subject to undergraduates at Manchester University. The books already available seemed to fall into two main categories: those that, although excellent in their time, had become rather out-dated and those that were aimed primarily at research workers. Hence my intention in producing this work has been to provide an up to date introduction to this rapidly expanding and exciting field.

I have aimed this book mainly at an audience of final year undergraduate or master students in any numerate discipline, as although it is mathematically oriented it does not demand a knowledge of any specific areas of advanced mathematics. Indeed all that is required of the student is an ability to think clearly and logically. Because I am not writing for the professional mathematician I have made considerable efforts to keep the notation as simple as possible, even at the expense of some loss of brevity. Many examples have been included to ensure understanding of the theory involved, and hints and solutions provided to the problems at the end of each chapter as an aid to those studying alone.

I am optimistic that this book will find an additional audience amongst professional operational research workers, for they too have need of a basic introduction to the subject. With this group particularly in mind I have extensively surveyed current developments in the theory of scheduling, and included in the bibliography sufficient references to guide them further into the literature. However I would stress that the emphasis here is more theoretical than practical, and that I have no intention of providing a "recipe book" for those faced with a particular scheduling problem.

The comments of many colleagues, friends, and students have been extremely helpful to me during the compilation of this book; in particular those of Professor B. W. Conolly and Dr. J. R. Walters who have read my various drafts and made many constructive criticisms and suggestions. The patience and skill of Kate Baker, who has undertaken all of the secretarial work, has been much appreciated. And finally it has indeed been a pleasure to work with my publisher, Ellis Horwood, his family and staff: their efficiency and tactful guidance have greatly smoothed the preparation of this book.

SIMON FRENCH
University of Manchester

1st March, 1981

Chapter 1

Scheduling Problems

1.1 INTRODUCTORY EXAMPLE

Algy, Bertie, Charles and Digby share a flat. Each Saturday they have four newspapers delivered: The *Financial Times*, the *Guardian*, the *Daily Express*, and the *Sun*. Being a small-minded creature of habit, each member of the flat insists on reading all the papers in his own particular order. Algy likes to begin with the *Financial Times* for 1 hour. Then he turns to the *Guardian* taking 30 minutes, glances at the *Daily Express* for 2 minutes and finishes with 5 minutes spent on the *Sun*. Bertie prefers to begin with the *Guardian* taking 1 hour 15 minutes; then he reads the *Daily Express* for 3 minutes, the *Financial Times* for 25 minutes, and the *Sun* for 10 minutes. Charles begins by reading the *Daily Express* for 5 minutes and follows this with the *Guardian* for 15 minutes, the *Financial Times* for 10 minutes, and the *Sun* for 30 minutes. Finally, Digby starts with the *Sun* taking 1 hour 30 minutes, before spending 1 minute each on the *Financial Times*, the *Guardian*, and the *Daily Express* in that order. Each is so insistent upon his particular reading order that he will wait for his next paper to be free rather than select another. Moreover, no one will release a paper until he has finished it. Given that Algy gets up at 8.30 a.m., Bertie and Charles at 8.45 a.m., and Digby at 9.30 a.m. and that they can manage to wash, shower, shave, and eat breakfast, while reading a newspaper, and given that each insists on reading all the newspapers before going out, what is the earliest time that the four of them can leave together for a day in the country?

The problem faced by Algy and his friends, namely in what order should they rotate the papers between themselves so that all the reading is finished as soon as possible, is typical of the scheduling problems that we shall be considering. Before describing the general structure of these problems and giving some examples that are, perhaps, more relevant to our modern society, it is worth examining this example further.

Let us write the data in a more compact form, as in Table 1.1. How might we tackle this problem? Perhaps it will be easiest to begin by explain-

Table 1.1—The data for the newspaper reading problem

Reader	Gets Up At	Reading Order And Times In Mins.			
Algy	8.30	F.T. (60)	G. (30)	D.E. (2)	S. (5)
Bertie	8.45	G. (75)	D.E. (3)	F.T. (25)	S. (10)
Charles	8.45	D.E. (5)	G. (15)	F.T. (10)	S. (30)
Digby	9.30	S. (90)	F.T. (1)	G. (1)	D.E. (1)

ing what is meant by a **reading schedule**. Quite simply it is a prescription of the order in which the papers rotate between readers. For instance, one possible schedule is shown in Table 1.2, where A, B, C, D denote Algy, Bertie, Charles, and Digby respectively. Thus Algy has the *Financial Times* first, before it passes to Digby, then to Charles, and finally to Bertie. Similarly the *Guardian* passes between Bertie, Charles, Algy, and Digby in that order. And so on.

Table 1.2—A possible reading schedule

Paper	Read by			
	1st	2nd	3rd	4th
F.T.	A	D	C	B
G.	B	C	A	D
D.E.	C	B	A	D
S.	D	A	C	B

We may work out how long this reading schedule will take by plotting a simple diagram called a **Gantt Diagram** (see Fig. 1.1). In this we plot four time-axes, one for each newspaper. Blocks are placed above the axes to indicate when and by whom particular papers are read. For instance, the block in the top left hand corner indicates that Algy reads the *Financial Times* from 8.30 to 9.30. To draw this diagram we have rotated the papers

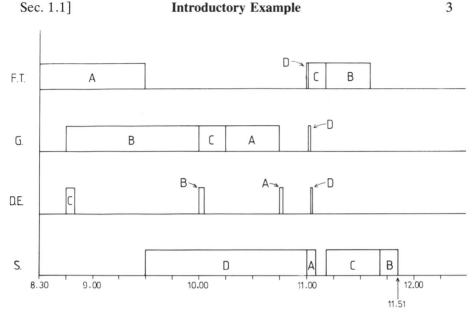

Fig.1.1—Gantt diagram for the schedule in Table 1.2.

in the order given by Table 1.2 with the restriction that each reader follows his desired reading order. This restriction means that for some of the time papers are left unread, even when there are people who are free and have not read them yet; they must remain unread until someone is ready to read them *next*. For instance, Bertie could have the *Financial Times* at 10.00 a.m., but he wants the *Daily Express* first and so leaves the *Financial Times*. Similarly the schedule is also responsible for idle time of the readers. Between 10.15 and 11.01 Charles waits for the *Financial Times*, which for all but the last minute is not being read, but Charles cannot have the paper until after Digby because of the schedule.

Clearly, from the Gantt diagram, the earliest that all four can go out together is 11.51 a.m. *if* they use this schedule. So the next question facing us is can we find them a better schedule, i.e. one that allows them to go out earlier. I am leaving that question for you to consider in the first set of problems. However, before attempting those, we should consider feasible and infeasible schedules.

In Table 1.2 I simply gave you a schedule and we saw in the Gantt diagram that this schedule would work; it is possible for Algy and his flatmates to pass the papers amongst themselves in this order. But suppose I had given you the schedule shown in Table 1.3. A little thought shows that this schedule will not work. Algy is offered the *Sun* first, but he does not want it until he has read the other three papers. He cannot have any of

Table 1.3—An infeasible schedule

Paper	Read by			
	1st	2nd	3rd	4th
F.T.	D	B	A	C
G.	D	C	B	A
D.E.	D	B	C	A
S.	A	D	C	B

these until Digby has finished with them and Digby will not start them until he has read the *Sun*, which he cannot have until Algy has read it. . . .

In scheduling theory the reading orders given in Table 1.1 are called the **technological constraints**. Any schedule which is compatible with these is called **feasible**. Thus Table 1.2 gives a feasible schedule. **Infeasible** schedules, such as that in Table 1.3, are incompatible with the technological constraints. Obviously, to be acceptable a solution to a scheduling problem must be feasible.

1.2 PROBLEMS

Please try these problems now before reading any further. It is true that I have given you little or no guidance on how they are to be solved. I have done so for good reason. Scheduling is a subject in which the problems look easy, if not trivial. They are, on the contrary, among the hardest in mathematics. You will not appreciate this without trying some for yourself. Solving them is relatively unimportant; I shall solve them for you shortly anyway. What is important is that you should discover their difficulty.

1. Is the following schedule feasible for Algy and his friends?

Paper	Read By			
	1st	2nd	3rd	4th
F.T.	C	D	B	A
G.	B	C	A	D
D.E.	B	C	A	D
S.	B	A	D	C

2. How many different schedules, feasible or infeasible, are there?

3. What is the earliest time that Algy and his friends can leave for the country?

4. Digby decides that the delights of a day in the country are not for him. Instead he will spend the morning in bed. Only when the others have left will he get up and read the papers. What is the earliest time that Algy, Bertie, and Charles may leave?

5. Whether or not you have solved Problems 3 and 4, consider how you would recognise the earliest possible departure time. Need you compare it explicitly with those of all the other feasible schedules, or can you tell without this process of complete enumeration of all the possibilities?

1.3 THE GENERAL JOB-SHOP SCHEDULING PROBLEM

If the theory of scheduling were simply concerned with the efficient reading of newspapers, then, of course, no one would study it. I began with that example so that you might meet and attempt to solve a scheduling problem unhindered by the definitions and notations that are usually required. The time has come to introduce these definitions and that notation. The terminology of scheduling theory arose in the processing and manufacturing industries. Thus we shall be talking about jobs and machines, even though in some cases the objects referred to bear little relation to either jobs or machines. For instance, in the example of the last section we shall see that Algy, Bertie, Charles and Digby are 'jobs', whilst the newspapers are 'machines'. However, that is anticipating. We begin by defining the **general job-shop problem**. Shortly we shall show that its structure fits many scheduling problems arising in business, computing, government, and the social services as well as those in industry.

We shall suppose that we have n **jobs** $\{J_1, J_2, \ldots J_n\}$ to be processed through m **machines** $\{M_1, M_2, \ldots M_m\}$. Some authors, particularly those writing on computer scheduling, refer to machines as **processors**. We shall suppose that each job must pass through each machine once and once only. The processing of a job on a machine is called an **operation**. We shall denote the operation on the ith job by the jth machine by o_{ij}. **Technological constraints** demand that each job should be processed through the machines in a particular order. For general job-shop problems there are no restrictions upon the form of the technological constraints. Each job has its own processing order and this may bear no relation to the processing order of any other job. However, an important special case arises when all the jobs share the same processing order. In such circumstances we say that we have a **flow-shop problem** (because the jobs *flow* between the machines in the same order). This distinction between job-shops and flow-shops should be made clear by the examples below.

Each operation o_{ij} takes a certain length of time, the **processing time**, to perform. We denote this by p_{ij}. By convention we include in p_{ij} any time required to adjust, or **set up**, the machine to process this job. We also include any time required to transport the job to the machine. We shall assume that the p_{ij} are fixed and known in advance. This brings us to an important restriction, which we shall make throughout this book. We shall assume that every numeric quantity is deterministic and known to the scheduler.

We shall also assume that the machines are always available, but we shall not necessarily assume the same for jobs. Some jobs may not become available until after the scheduling period has started. We shall denote by r_i the **ready time** or **release date** of the ith job, i.e. the time at which J_i becomes available for processing.

The general problem is to find a sequence, in which the jobs pass between the machines, which is

(a) compatible with the technological constraints, i.e. a feasible schedule, and
(b) optimal with respect to some criterion of performance.

We shall discuss possible criteria of performance shortly; first we shall consider some particular examples.

Industrial examples

Any manufacturing firm not engaged in mass production of a single item will have scheduling problems at least similar to that of the job-shop. Each product will have its own route through the various work areas and machines of the factory. In the clothing industry different styles have different requirements in cutting, sewing, pressing and packing. In steel mills each size of rod or girder passes through the set of rollers in its own particular order and with its own particular temperature and pressure settings. In the printing industry the time spent in typesetting a book will depend on its length, number of illustrations, and so on; the time spent in the actual printing will depend on both its length and the number printed; the time spent on binding will depend on the number printed; and lastly the time spent in packaging will depend both on the number printed and the book's size. Thus a printer who must schedule the production of various books through his typesetting, printing, binding and packaging departments faces a four-machine flow-shop problem; for each department is a machine and all the jobs, i.e. the books, flow from typesetting to printing to binding to packaging.

The objectives in scheduling will vary from firm to firm and often from day to day. Perhaps the aim would be to maintain an equal level of activity in all departments so that expensive skills and machines were seldom idle.

Perhaps it would be to achieve certain contractual target dates or simply to finish all the work as soon as possible. These and some other objectives will be discussed in detail in section 1.5 and in all of Chapter 2.

Algy and friends

This is a four job, four machine problem. The jobs—Algy, Bertie, Charles, and Digby—must be scheduled through the four machines—the *Financial Times*, the *Guardian*, the *Daily Express*, and the *Sun*—in order to minimise the completion time at which the processing, here reading, is finished There are ready times, namely, the times at which Algy, etc. get up. Note that in this example the technological constraints do not demand the flat-mates read the papers in the same order. Because of this we have a job-shop problem. Had the constraints demanded that each read the papers in the same order, e.g. *Sun, Guardian, Daily Express* and, finally, the *Financial Times*, we would have had a flow-shop problem.

Aircraft queuing up to land

This is an *n* job, one machine problem, if we assume, as we do, that the number of aircraft arriving in a day is known. The aircraft are the jobs and the runway is the machine. Each aircraft has a ready time, namely the earliest time at which it can get to the airport's air-space and be ready to land. The objective here might be to minimise the average waiting time of an aircraft before it can land. Perhaps this average should be weighted by the number of passengers on each plane. Obviously in real life this problem models only part of the air traffic controllers' difficulties. Planes must take off too. Also it ignores the uncertainty inherent in the situation. Aircraft suffer unpredictable delays and, moreover, the time taken to land, i.e. the processing time of each aircraft, will depend on the weather.

Treatment of patients in a hospital

Again we ignore all randomness. Suppose we have *n* patients who must be treated by a consultant surgeon. Then each must be

M_1—seen in the out-patients' department,
M_2—received in the surgical ward, prepared for the operation, etc.,
M_3—be operated on.
M_4—recover, we hope, in the surgical ward.

Thus each patient (job) must be processed through each of four 'machines', M_1, M_2, M_3 and M_4 above. It is perhaps confusing that only one of the operations of processing a patient through a 'machine' is called a surgical operation, but in scheduling theory all are operations. Since each patient must clearly 'flow' through the 'machines' in the order M_1, M_2, M_3, M_4, this is a flow-shop problem. The objective here might be stated as: treat all patients in the shortest possible time, whilst giving priority to the most ill.

Other scheduling problems

Given these examples you should have no difficulty in fitting other practical problems into the general job-shop structure. For instance, the following all fall into this pattern:

    (a) the scheduling of different programs on a computer;
    (b) the processing of different batches of crude oil at a refinery;
    (c) the repair of cars in a garage;
    (d) the manufacture of paints of different colours.

1.4 ASSUMPTIONS

For the major part of this book we shall be making a number of assumptions about the structure of our scheduling problems. Some were mentioned explicitly above; others were implicit. Here we list all the assumptions both for further emphasis and for easy reference. Why we choose to make these assumptions, which are often quite restrictive, is discussed in section 1.7.

1. Each job is an entity. Although the job is composed of distinct operations, no two operations of the same job may be processed simultaneously. Thus we exclude from our discussion certain practical problems, e.g. those in which components are manufactured simultaneously prior to assembly into the finished product.

2. No pre-emption. Each operation, once started, must be completed before another operation may be started on that machine.

3. Each job has m distinct operations, one on each machine. We do not allow for the possibility that a job might require processing twice on the same machine. Equally, we insist that each job is processed on every machine; it may not skip one or more machines. Note that this latter constraint is *not* illusory. Although we could say that a job which skips a machine is processed upon that machine for zero time, we would still have a problem: where in the job's processing sequence should this null operation be placed. Because we do not allow pre-emption, the job could be delayed waiting for a machine which is not in fact needed.

4. No cancellation. Each job must be processed to completion.

5. The processing times are independent of the schedule. In particular we are assuming two things here. Firstly, each set-up time is sequence-independent, i.e. the time taken to adjust a machine for a job is independent of the job last processed. Secondly, the times to move jobs between machines are negligible.

6. *In-process inventory is allowed;* i.e. jobs may wait for their next machine to be free. This is not a trivial assumption. In some problems processing of jobs must be continuous from operation to operation. In steel mills one literally has to strike while the iron is hot.

7. *There is only one of each type of machine.* We do not allow that there might be a choice of machines in the processing of a job. This assumption eliminates, amongst others, the case where certain machines have been duplicated to avoid bottlenecks.

8. *Machines may be idle.*

9. *No machine may process more than one operation at a time.*†

10. *Machines never breakdown and are available throughout the scheduling period.*

11. *The technological constraints are known in advance and are immutable.*

12. *There is no randomness.* In particular

(a) the number of jobs is known and fixed;
(b) the number of machines is known and fixed;
(c) the processing times are known and fixed;
(d) the ready times are known and fixed;
(e) all other quantities needed to define a particular problem are known and fixed.

Occasionally we shall relax one or two of these assumptions in specific examples. When we do, we shall state so explicitly.

1.5 PERFORMANCE MEASURES

It is not easy to state our objectives in scheduling. They are numerous, complex, and often conflicting. Mellor (1966) lists 27 distinct scheduling goals, albeit in a slightly broader context. In addition, the mathematics of our problem can be extremely difficult with even the simplest of objectives. So we shall not try to be exhaustive. Instead we shall indicate in general terms a few of the criteria by which we might judge our success. These criteria will be sufficient to motivate the mathematical definitions of the performance measures that we shall use. Whether these abstractions are so simple-minded that our theory ceases to have practical relevance is discussed in section 1.7.

Obviously we should prefer to keep promised delivery dates. Otherwise good-will would surely be lost and there might be financial penalties as well. Indeed one may argue cogently that a delivery date that is not

---

†To a mathematician assumptions 7 and 9 are identical; for mathematically two of the same type of machine is equivalent to a single machine capable of processing two jobs simultaneously. However, we make both assumptions explicitly for those with a less mathematical view of the world.

enforced by some penalty is not truly a delivery date. We should also try to minimise the overall length of the scheduling period, because once all the jobs have been completed the machines may be released for other tasks. We should try to minimise the time for which the machines are idle; idle machines mean idle capital investment. We should try to minimise inventory costs and by these we do not mean just the cost of storing the finished product. There are also the costs of storing the raw materials and any partially processed jobs that must wait between machines. We might try to ensure a uniform rate of activity throughout the scheduling period so that demands for labour and power are stable. Conversely it might be desirable to concentrate the activity into periods when either labour or power is particularly cheap.

Before we can define performance measures in precise mathematical terms, we need some more definitions and notation. Remember that r_i and p_{ij} are respectively the ready time and processing times of job J_i.

d_i is the **due date**, i.e. the promised delivery date of J_i. It is the time by which ideally we would like to have completed J_i.

a_i is the **allowance** for J_i. It is the period allowed for processing between the ready time and the due date: $a_i = d_i - r_i$.

W_{ik} is the **waiting time** of J_i preceding its kth operation. By kth operation we do not mean the one performed on M_k (although it may be), but the one that comes kth in order of processing. Thus, if the technological constraints demand that J_i is processed through the machines in the order $M_{j(1)}, M_{j(2)}, M_{j(3)} \ldots, M_{j(m)}$, the kth operation is $o_{ij(k)}$, the one performed on $M_{j(k)}$. So W_{ik} is the time that elapses between the completion of J_i on $M_{j(k-1)}$ (or r_i if $k = 1$) and the start of processing on M_k.

W_i is the **total waiting time** of J_i. Clearly $W_i = \sum_{k=1}^{m} W_{ik}$

C_i is the **completion time** of J_i, i.e. the time at which processing of J_i finishes. We have the equality: $C_i = r_i + \sum_{k=1}^{m} (W_{ik} + p_{ij(k)})$.

F_i is the **flow time** of J_i. This is defined to be the time that J_i spends in the workshop. Thus $F_i = C_i - r_i$.

L_i is the **lateness** of J_i. This is simply the difference between its completion time and its due date: $L_i = C_i - d_i$. Note that when a job is early, i.e. when it completes before its due date, L_i is negative. It is often more use to have a variable which, unlike lateness, only takes non-zero values when a job is **tardy**, i.e. when it completes after its due date. Hence we also define the tardiness and, to be comprehensive, the earliness of a job.

T_i is the **tardiness** of J_i: $T_i = \max\{L_i, 0\}$.

E_i is the **earliness** of J_i: $E_i = \max\{-L_i, 0\}$.

These quantities for a typical job are illustrated on the Gantt diagram shown in Fig. 1.2. There is a possibility of confusion in some of this terminology. English allows the noun 'time' two distinct meanings. It may be used to refer to an instant or to an interval. Thus ready time and completion time both refer to instants in time; whereas processing time, waiting time, and flow time refer to intervals. In remarking upon this ambiguity we hope to avoid possible confusion. In any case, the context usually ensures that the meaning is clear.

We shall often need to discuss the maximum or the mean of these quantities and it will help to have a compact notation for doing this. Let X_i be any quantity relating to J_i. Then we let $\bar{X} = 1/n \sum_{i=1}^{n} X_i$, the average over all the jobs, and $X_{max} = \max\{X_1, X_2, \ldots, X_n\}$, the maximum over all the jobs. For instance, $\bar{F}$ is the mean flow time and C_{max} is the maximum completion time.

Next we define the **idle time** on machine M_j by $I_j = C_{max} - \sum_{i=1}^{n} p_{ij}$. In order to see that this definition makes sense, note that C_{max} is the time

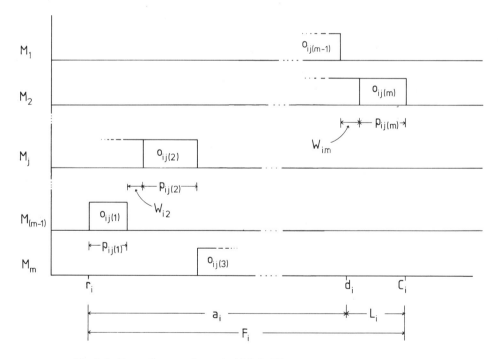

Fig. 1.2 Gantt diagram of a typical job J_i. The processing order given by the technological constraints is $(M_{(m-1)}, M_j, M_m, \ldots, M_1, M_2)$. The waiting times W_{i1}, W_{i3} are zero; W_{i2} and W_{im} are non-zero as shown. For this job $T_i = L_i$ and $E_i = 0$, since the job is completed after its due date.

when all processing ceases and $\Sigma_{i=1}^{n} p_{ij}$ is the total processing time on machine M_j. Their difference gives the period for which M_j is idle.

Finally, we introduce some variables which count the number of jobs in various states at any given time. We let

$N_W(t)$ the number of jobs waiting between machines or not ready for processing at time t;

$N_p(t)$ be the number of jobs actually being processed at time t;

$N_c(t)$ be the number of jobs completed by time t; and

$N_u(t)$ be the number of jobs still to be completed by time t.

Clearly we have the following identities:

$$\left.\begin{array}{l} N_W(t) + N_p(t) + N_c(t) = n \\ N_W(t) + N_p(t) \qquad\quad = N_u(t) \end{array}\right\} \text{ for all } t,$$

and

$$\begin{array}{l} N_u(0) \quad\; = n, \\ N_u(C_{\max}) = 0. \end{array}$$

We shall again use an over-bar to indicate an average for these quantities, the average being, of course, defined relative to the time period, e.g.

$$\bar{N}_u = \frac{1}{C_{\max}} \int_0^{C_{\max}} N_u(t)\, \mathrm{d}t.$$

Now we are in a position to define some measures of performance.

Criteria based upon completion times

The main criteria in this category are $F_{\max}$, $C_{\max}$, $\bar{F}$ and $\bar{C}$. Minimising $F_{\max}$, the maximum flow time, is essentially saying that a schedule's cost is directly related to its longest job. Minimising $C_{\max}$, the maximum completion time, says that the cost of a schedule depends on how long the processing system is devoted to the entire set of jobs. Note that in the case where all the ready times are zero $C_{\max}$ and $F_{\max}$ are identical. However, when there are non-zero ready times, $C_{\max}$ and $F_{\max}$ are quite distinct. Indeed, if one job has an extremely late ready time, it may easily happen that the job with the *shortest* flow-time completes at $C_{\max}$. It is appropriate to mention here that $C_{\max}$ is also called the **total production time** or the **make-span**. Minimising $\bar{F}$, the mean flow-time, implies that a schedule's cost is directly related to the average time it takes to process a single job. We shall find that minimising $\bar{C}$, the mean completion time, is equivalent to minimising $\bar{F}$; i.e. a schedule which attains the minimum $\bar{C}$ also attains the minimum $\bar{F}$ and vice versa (see Theorem 2.2). It may seem strange that $F_{\max}$ and $C_{\max}$ are quite distinct measures of performance, whereas $\bar{F}$ and $\bar{C}$ are essentially the same. The reason for this is quite simple. The operation of taking the maximum of a set of numbers has different properties to that of taking the mean. For a further discussion of this, see Theorem 2.2 and the remarks that follow it.

Although we shall seldom do so, some authors have considered weighted measures, which recognise that some jobs are more important than others. Thus they suggest that we minimise weighted averages,

$$\sum_{i=1}^{n} \alpha_i C_i \quad \text{or} \quad \sum_{i=1}^{n} \beta_i F_i$$

where $\alpha_1, \alpha_2, \ldots, \alpha_n$ and $\beta_1, \beta_2, \ldots, \beta_n$ are weighting factors usually summing to 1.

Criteria based upon due dates

Since the cost of a schedule is usually related to how much we miss target dates by, obvious measures of performance are $\bar{L}$, L_{max}, $\bar{T}$ and T_{max}; i.e. the mean lateness, the maximum lateness, the mean tardiness, and the maximum tardiness respectively. Minimising either $\bar{L}$ or L_{max} is appropriate when there is a positive reward for completing a job early, and that reward is larger the earlier a job is. Minimising either $\bar{T}$ or T_{max} is appropriate when early jobs bring no reward; there are only the penalties incurred for late jobs.

Sometimes the penalty incurred by a late job does not depend on how late it is; a job that completes a minute late might just as well be a century late. For instance, if an aircraft is scheduled to land at a time after which it will have exhausted its fuel, then the results are just as catastrophic whatever the scheduled landing time. In such cases, a reasonable objective would be to minimise n_T, **the number of tardy jobs**, i.e. the number of jobs that complete after their due dates.

Criteria based upon the inventory and utilisation costs

Here we might wish to minimise $\bar{N}_w$, the mean number of jobs waiting for machines, or $\bar{N}_u$, the mean number of unfinished jobs. Both of these measures are roughly related to the in-process inventory costs. We might be more concerned with minimising $\bar{N}_c$, the mean number of completed jobs, because doing this will, in general, reduce the inventory costs of the finished goods. If our aim is to ensure the most efficient use of the machines, then we might choose to maximise $\bar{N}_p$, the mean number of jobs actually being processed at any time. Alternatively we might seek the objective of efficient machine use by minimising $\bar{I}$ or I_{max}, the mean or the maximum machine idle time. (N.B. For idle times the mean and the maximum are taken over the machines, not the jobs.)

Finally, we note a classification of performance measures into those that are regular and those that are not. A **regular measure** R is simply one that is non-decreasing in the completion times. Thus R is a function of $C_1, C_2, \ldots, C_n$ such that

$$C_1 \le C_1', \quad C_2 \le C_2', \ldots, C_n \le C_n'$$

together $\Rightarrow R(C_1, C_2, \ldots, C_n) \le R(C_1', C_2', \ldots, C_n')$.

The rationale underlying this definition is as follows: Suppose that we have two schedules such that under the first all the jobs complete no later than they do under the second. Then under a regular performance measure the first schedule is at least as good as the second. Note that we seek to minimise a regular measure of performance.

$\bar{C}, C_{max}, \bar{F}, F_{max}, \bar{L}, L_{max}, \bar{T}, T_{max}$ and n_T are all regular measures of performance, as a number of simple arguments show. For instance, consider the performance measure C_{max}. Here we have

$$\begin{aligned} R\{C_1, C_2, \ldots, C_n\} &= C_{max} \\ &= \max\{C_1, C_2, \ldots, C_n\}. \end{aligned}$$

Let $C_1 \le C_1', C_2 \le C_2', \ldots, C_n \le C_n'$. Then

$$\begin{aligned} R\{C_1, C_2, \ldots, C_n\} &= \max\{C_1, C_2, \ldots, C_n\} \\ &\le \max\{C_1', C_2', \ldots, C_n'\} \\ &= R\{C_1', C_2', \ldots, C_n'\}. \end{aligned}$$

Hence C_{max} is a regular performance measure.

1.6 CLASSIFICATION OF SCHEDULING PROBLEMS

It will be convenient to have a simple notation to represent types of job-shop problems. We shall classify problems according to four parameters: $n/m/A/B$.

n is the number of jobs.

m is the number of machines.

A describes the flow pattern or discipline within the machine shop. When $m = 1$, A is left blank. A may be

    F for the flow-shop case, i.e. the machine order for all jobs is the same.

    P for the permutation flow-shop case. Here not only is the machine order the same for all jobs, but now we also restrict the search to schedules in which the job order is the same for each machine. Thus a schedule is completely specified by a single permutation of the numbers $1, 2, \ldots, n$, giving the order in which the jobs are processed on each and every machine.

    G the general job-shop case where there are no restrictions on the form of the technological constraints.

    B   describes the performance measure by which the schedule is to be evaluated. It may take, for instance, any of the forms discussed in the previous section.

As an example: $n/2/F/C_{max}$ is the n job, 2 machine, flow-shop problem where the aim is to minimise make-span.

    In using four parameters we follow Conway, Maxwell, and Miller (1967). Other authors, notably Lenstra (1977), Rinnooy Kan (1976), and Graham *et al.* (1979), introduce further parameters, but their discussions extend over a much larger family of problems than we shall be considering.

1.7 REAL SCHEDULING PROBLEMS AND THE MATHEMATICS OF THE JOB-SHOP

One need not have encountered scheduling problems in practice to realise that they are vastly more complex than those of the job-shop as we have defined them. Few obey many, much less all, of the assumptions that we have made. The cost of a schedule is seldom well represented by a function as simple as $\bar{C}$ or T_{max}. It might be expected, therefore, that there would be little practical relevance of the theory which we will develop. In fact, our analysis is not at all irrelevant and it would be wise to pause and consider why.

    Firstly let us examine assumptions 1 to 11 of section 1.4. (Assumption 12 differs in nature from the others and will be discussed separately.) These assumptions are undoubtedly restrictive. They limit the structure of the job-shop problem greatly. They define quite precisely which routings of jobs are allowable and which are not; what the capacities and availabilities of the machines are; etc. It is very easy to imagine practically occurring problems where some of these restrictions are totally inappropriate. So it is comforting to find that these assumptions are not necessary to the development of a mathematical theory of scheduling; rather they are typical of the assumptions which we might make. We could have chosen another set of assumptions, and so defined another family of scheduling problems, and then developed a theory of scheduling for the family. Had we done so, we would have encountered the same difficulties, the same forms of argument, the same mathematical techniques, and essentially the same results as we shall here. The reason for choosing the job-shop family is that it leads to a presentation of the theory which is particularly coherent and, furthermore, is not encumbered with a confusion of caveats and provisos needed to cover special cases. Once the job-shop family has been studied, it will be an easy matter to follow developments of the theory in other contexts. To help in this, Chapter 12 is a brief survey of such developments; it defines many other families of scheduling problems and references the literature appropriate to them.

    Assumption 12 and, to a small extent, Assumption 10, stand quite distinct from the rest. They confine attention to non-random problems, that is problems in which *all* numerical quantities are known and fixed in advance. There is no uncertainty. Because the number of jobs and their ready times are known and fixed, we call our problems **static**. Because the processing times and all other parameters are known and fixed, we call our problems **deterministic**. Problems in which jobs arrive randomly over a period of time are called **dynamic**. Problems in which the processing times, etc. are uncertain are called **stochastic**. It may be argued that all practical scheduling problems are both dynamic and stochastic, if for no other reason than that all quantities are subject to some uncertainty. In fact, in many problems the randomness is quite obvious. For instance, it may be impossible to predict exactly when jobs will become available for processing, e.g. aircraft arriving at an airport's airspace; it may be impossible to predict processing times exactly, e.g. during routine maintenance it will not be known which parts have to be replaced until they have been examined and that examination is one of the operations of the maintenance process; it may be impossible to predict the availability of machines, for some may have significant breakdown rates; and so on. Given that most problems have dynamic and stochastic elements, why do we confine ourselves to static, deterministic cases?

To begin with, there are problems in which any randomness is quite clearly insignificant, i.e. the uncertainty in the various quantities is several orders of magnitude less than the quantities themselves. Indeed, as microprocessors and industrial robots enter production lines, we may expect a greater degree of certainty in processing times. Secondly, we cannot study the dynamic and stochastic families of problems until we have understood the static, deterministic family. An awareness of the techniques for scheduling jobs when there is no uncertainty involved will point us towards the solution of stochastic problems. Finally, there is a rather negative reason for omitting the study of dynamic and stochastic problems from this book. To have included it would have required a deeper knowledge of probability theory and statistics than appropriate to an introductory text.[†] The literature of the dynamic stochastic problem is briefly surveyed and referenced in Chapter 12.

So we shall accept all twelve assumptions of section 1.4 without further question, and that leaves us just one more point to discuss here: is our choice of performance measures too limited to allow the representation of the scheduling goals that arise in practice. The first point to note is that in using a performance measure such as $\bar{T}$ we are not assuming that the cost of a schedule is directly proportional to $\bar{T}$, i.e. that the cost is a positive linear

---

[†]There is a fourth reason for omitting a study of dynamic and stochastic problems. This book may be looked upon as an introduction not to the theory of scheduling, but rather to the theory of combinatorial optimisation. Stochastic problems are not usually considered under the heading of combinatorial optimisation; they involve different branches of mathematics.

function of $\bar{T}$. What we are assuming is that the cost is an increasing function of $\bar{T}$, i.e. a function such that the cost increases when and only when $\bar{T}$ increases; the increases in cost and $\bar{T}$ need not be proportional. Under this condition minimising $\bar{T}$ minimises the cost. Thus in restricting the choice of performance measures to those listed in section 1.5, we are not restricting the form of cost function quite as much as we might think. Nonetheless, we have limited our choice and we should consider the implications of this.

We are limiting ourselves not because it is wrong to want to minimise, say C_{max}, but because in any real problem we would also want to minimise $\bar{L}, \bar{I}$, etc. The total cost of a schedule is a complex combination of processing costs, inventory costs, machine idle-time costs, and lateness penalty costs. In other words, each of our performance measures represents only a component of the total cost. A schedule which minimises a component cost may be very poor in terms of the total cost.

In fact, our study will lose little by its failure to consider total costs. We shall discover that even with simple performance measures scheduling problems can be extremely difficult. Their solution usually requires heuristic or approximate methods (see Chapters 10 and 11), and these methods may easily be modified to minimise total costs. Moreover, the insight necessary to enable such modifications will come from our earlier work on problems with simple performance measures.

Gupta and Dudek (1971), and Rinnooy Kan (1976, pp. 24–28) discuss the form of the total scheduling cost. Also Van Wasenhove and Gelders (1980) have introduced the concept of efficiency, i.e. Pareto optimality, into the field of scheduling, and we shall study this briefly in Chapter 4.

1.8 ALGY, BERTIE, CHARLES, AND DIGBY'S FLAT REVISITED

Please do not read this section until you have tried Problems 1.2.

Usually I shall only sketch the solutions to the problems that I set (see pp. 207–229). However, the problems concerning Algy and his friends are so important to the development of the theory that we should pause and examine their solution in some detail.

1. Is the given schedule feasible? The short answer is no, and we may discover this in a number of ways. We might try to draw a Gantt diagram of the schedule and find that it is impossible to place some of the blocks without conflicting with either the schedule or the technological constraints. Alternatively we might produce an argument similar to, but more involved than, that by which we showed the schedule in Table 1.3 to be infeasible. It will be agreed though that neither of these methods is particularly straightforward and, moreover, that the thought of extending either to larger problems is awesome. What we need is a simple scheme for checking the schedule operation by operation until either a conflict with the technological constraints is found or the whole schedule is shown to be feasible. The following illustrates such a scheme.

Firstly we write the schedule and the technological constraints side by side as below. You should ignore the 'circles' for the time being. We shall imagine that we are operating the schedule. We shall assign papers to readers as instructed. As the papers are read we shall circle the operations to indicate that they are completed and pass the papers to their next readers. Either we shall meet an impasse or we shall show that the schedule is feasible. (Note that I have subscripted the circles in the order in which they are entered.)

Schedule				Technological Constraints				
	Read By							
Paper	**1st**	**2nd**	**3rd**	**4th**	**Reader**	**Order of Reading Papers**		
F.T.	$\textcircled{C}_5$	D	B	A	A	F.T.	G.	D.E.	S.
G.	$\textcircled{B}_1$	$\textcircled{C}_4$	A	D	B	$\textcircled{G.}_1$	$\textcircled{D.E.}_2$	F.T.	S.
D.E.	$\textcircled{B}_2$	$\textcircled{C}_3$	A	D	C	$\textcircled{D.E.}_3$	$\textcircled{G.}_4$	$\textcircled{F.T.}_5$	S.
S.	B	A	D	C	D	S.	F.T.	G.	D.E.

We begin in the top left hand corner of the schedule. Charles is given the *Financial Times*, but will not read it until he has read the *Daily Express* and the *Guardian*. So we must leave this operation uncompleted and uncircled. Proceeding down the schedule, Bertie is given the *Guardian* and we see from the technological constraints that he is immediately ready to read it. So we circle this operation both in the schedule and in the technological constraints. Next Bertie is given the *Daily Express* and we see that, now he has read the *Guardian*, he is immediately ready to read it. So this is the second operation to be circled. The *Sun* we see is also assigned to Bertie, but he is not ready to read it so we leave this operation uncircled. We continue by returning to the top line of the schedule and by repeatedly working down the schedule, checking the leftmost uncircled operation in each line to see if it may be performed. Thus we show that Charles may read the *Daily Express*, the *Guardian*, and the *Financial Times* without any conflict with the technological constraints. The position is now as shown with five operations circled and we can see quite clearly that an impasse has been reached. Each paper is assigned to a reader who does not wish to read it yet. Hence the schedule is quite clearly infeasible.

It will transpire that we need to check for feasibility only very infrequently since the algorithms and methods that we shall study are designed so that they cannot produce infeasible schedules. For this reason I shall not write out the steps of the checking scheme explicitly; they should be clear enough from the example anyway. If you need to check a schedule for a more conventional problem based upon jobs and machines, you should have no difficulty in translating the method from the present context; just remember that here Algy and friends are the jobs, while the papers are the machines.

2. How many different schedules, feasible or infeasible, are there? A schedule for the general $n/m/G/B$ job-shop problem consists of m permutations of the n jobs. Each permutation gives the processing sequence of jobs on a particular machine. Now there are $n!$ different permutations of n objects and, since each of the m permutations may be as different as we please from the rest, it follows that the total number of schedules is $(n!)^m$. In the problem facing Algy and his friends $n = 4$ and $m = 4$. So the total number of schedules is $(4!)^4 = 331,776$.

It is worthwhile pausing to consider the implications of this rather startling number. Here we have a very small problem: only 4 'machines' and 4 'jobs'. Yet the number of possible contenders for the solution is enormous. We cannot hope to solve the problem by the simple expedient of listing all the possible schedules, eliminating the infeasible, and selecting the best of those remaining. To be fair, we might be able to check through the 331,776 possibilities in a reasonable time on a fast computer. If 1000 schedules were checked each second (which would be very fast), the computer would solve the problem in just over $5\frac{1}{2}$ minutes. But suppose that a guest were staying in the flat so that there were 5 readers. The number of schedules would now be $(5!)^4 = 2.1 \times 10^8$ and the computer would take over 57 hours to solve this new problem! The very size of these numbers indicates the very great difficulty of scheduling problems. To have any chance at all of solving them we must use subtlety. But even with the most subtle methods available we shall discover that some problems defy practical solution; to solve them would literally take centuries.

3. What is the earliest time at which Algy and his friends may leave for the country? Perhaps the easiest way for us to approach this problem is to look back at the schedule given in Table 1.2 and see if we can improve upon it in any obvious way. Looking at the Gantt diagram (Fig. 1.1) and, in particular the row for the *Sun*, we see that it is the *Sun* that is finished last. Moreover,

Table 1.4—An improved reading schedule

| | | Read By | | |
Paper	1st	2nd	3rd	4th
F.T.	A	B	C	D
G.	B	C	A	D
D.E.	C	B	A	D
S.	D	A	C	B

it is left unread between 11.05 when Algy finishes it and 11.11 when Charles is ready for it. Thus 6 minutes are apparently wasted. Is there another schedule which does not waste this time, one which ensures that the *Sun* is read continuously? Well yes, there is. Consider Table 1.4. This has the Gantt diagram shown in Fig. 1.3. Note that now the *Sun* is read continuously and that it is the last paper to be finished. Thus under this schedule the earliest time at which they can leave for the country is 11.45. Moreover, a little thought will convince most people that this is an optimal schedule. Everybody starts reading as soon as they can and once started the *Sun* is read without interruption. There seems to be no slack left in the

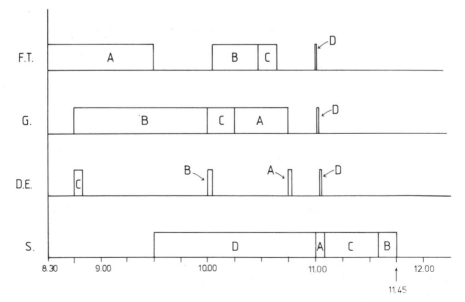

Fig. 1.3—Gantt diagram for the schedule in Table 1.4.

Table 1.5—A reading schedule showing still more improvement

Paper		Read By		
	1st	**2nd**	**3rd**	**4th**
F.T.	C	A	B	D
G.	C	B	A	D
D.E.	C	B	A	D
S.	C	D	B	A

system. But there is. Consider the schedule in Table 1.5. This schedule leads to the Gantt diagram shown in Fig. 1.4 and we see that all reading is now completed by 11.30, 15 minutes earlier than allowed by the schedule in Table 1.4. So that schedule is clearly not optimal. How has this improvement been achieved? Compare the rows for the *Sun* in the two Gantt diagrams (Figs 1.3 and 1.4). What we have done is 'leap-frogged' the block for Charles over those for Digby and Algy. Because Charles can be ready for the *Sun* at 9.15, if he is allowed the other papers as he wants them, we gain 15 minutes. Moreover, it is possible to schedule the other readers, Algy, Bertie, and Digby so that this gain is not lost. The moral of all this is that in scheduling you often gain overall by not starting a job on a

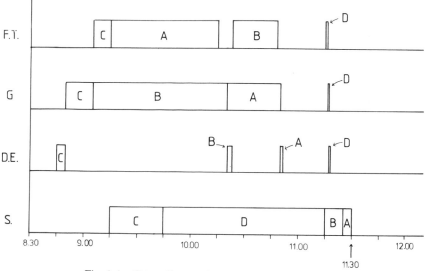

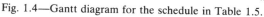

Fig. 1.4—Gantt diagram for the schedule in Table 1.5.

machine as soon as you might. Here Algy and Bertie wait for the *Financial Times* and *Guardian* respectively. They could snatch up these papers as soon as they get up, but their patience is a virtue rewarded.

It turns out that the schedule in Table 1.5 is optimal; no other schedule allows them to leave the flat earlier. To see this we consider four mutually exclusive possibilities: Algy reads the *Sun* before anyone else; Bertie does; Charles does; or Digby does. At the earliest Algy can be ready to read the *Sun* at 10.02. (Check this from Table 1.1.) The earliest times at which Bertie, Charles and Digby can be ready are 10.28, 9.15, and 9.30 respectively. Thus, if we assume that, once started, the *Sun* is read continuously taking 2 hr. 15 min. in total, then the earliest time at which all reading can finish in the four cases is 12.17, 12.43, 11.30 and 11.45 respectively. Note that these are lower bounds on the completion times. For instance, a schedule which gives Algy the *Sun* first might not finish at 12.17 either because other papers continue to be read after the *Sun* is finished or because it is not possible for the *Sun* to be read continuously. So the earliest possible time for any schedule to finish is min{12.17, 12.43, 11.30, 11.45} = 11.30. Table 1.5 gives a schedule completing at 11.30; it must, therefore, be optimal.

The structure of the above argument deserves special emphasis for it will be developed into a powerful solution technique known as branch and bound (see Chapter 7). We had a particular schedule which completed finally at a known time. To show that this schedule was optimal, we considered all possible schedules and divided them into four disjoint classes. We worked out for each class the earliest that any schedule within that class could complete, i.e. we found a lower bound appropriate to each of these classes. We then noted that our given schedule completed at the lowest of the lower bounds. Thus no other schedule could complete before it and so it had to be an optimal schedule.

4. What is the earliest time at which Algy, Bertie and Charles may leave without Digby? 11.03. I leave with you both the problem of finding a schedule to achieve this and the problem of showing such a schedule to be optimal. However, I will give you one hint. Use a bounding argument like that above except that you should consider who is first to read the *Guardian*, not the *Sun*.

5. How does one prove a schedule to be optimal? Need one resort to complete enumeration? For the particular scheduling problem facing Algy and friends we now know that complete enumeration is unnecessary. However, the solution of Problem 1.2.3 involved a certain element of luck, or rather relied on knowing the answer before we started. No straightforward logical argument led to the schedule in Table 1.5. I just produced it rather like a magician pulling a rabbit from a hat. Moreover, the bounding argument that I used to show optimality relied heavily on the structure of this particular problem, namely that the optimal schedule allowed the *Sun* to be read continuously. (Why is this particular feature important to the argument?) In short, I have been able to solve this problem for you simply because I set it. So the question remains: in general is it necessary to use complete enumeration to solve scheduling problems. The answer is the rest of this book.

1.9 GENERAL READING

Scheduling theory, as I have said, is a surprisingly difficult subject. Consequently there are few elementary introductions. Conway, Maxwell, and Miller (1967) is, perhaps, the simplest reference with Baker (1974) a close second. The papers of Mellor (1966) and Bakshi and Arora (1969) introduce the job-shop scheduling problem, but then go little further. There are several collections of papers available: notably Muth and Thompson (1963), Elmaghraby (1973), Coffman (1976), and a special issue of *Operations Research* (1978, Volume 26, No. 1). At a more advanced level are the monographs of Lenstra (1977) and Rinnooy Kan (1976). Graham *et al.* (1979) give an excellent, but mathematically terse, survey of the known results in scheduling theory as they stood at the end of 1977. Klee (1980) reviews the whole area of combinatorial optimisation, of which scheduling theory forms a part.

1.10 PROBLEMS

1. Choose a practical scheduling problem that you know about from your own experience. Try to fit it into the structure of the general job-shop. Consider carefully whether your problem obeys each assumption given in section 1.4 and discuss carefully what performance criterion is appropriate.

2. Show that for any job, J_i:

$$L_i = F_i - a_i = C_i - r_i - a_i = C_i - d_i$$

and hence deduce that:

$$\bar{L} = \bar{F} - \bar{a} = \bar{C} - \bar{r} - \bar{a} = \bar{C} - \bar{d}.$$

3. Show that $L_i = T_i - E_i$.

4. (i) Show that $\bar{F}$, T_{max}, and n_T are all regular measures of performance.

   (ii) Show that minimising $\bar{E}$ or E_{max} does not correspond to a regular measure. Why might it be sensible to want to minimise $\bar{E}$ or E_{max}?

5. Is the following schedule feasible for Algy and his friends?

Paper	Read By			
	1st	2nd	3rd	4th
F.T.	A	C	B	D
G.	C	A	B	D
D.E.	C	A	B	D
S.	A	D	C	B

Chapter 2

Optimality of Schedules

2.1 INTRODUCTION

Our problem in scheduling is to timetable the processing of jobs by machines so that some measure of performance achieves its optimal value. However, before we approach this directly, there are two pertinent questions which should be answered.

(1) For any particular measure of performance does an optimal schedule exist?

(2) Given that we have found a schedule that is optimal with respect to one measure, how does it fare against a second measure?

The first question may be a little perplexing, unless you are fluent enough in the language of mathematics to be aware of greatest lower bounds (infima), compact sets, etc. We shall avoid such terms, but the ideas are as follows. Suppose I asked you to find the smallest number strictly greater than 2. A little thought shows that this apparently simple task is impossible. If you give me the number 2, then I reply that 2 is not strictly greater than 2. If you give me a slightly greater number, say $(2 + \varepsilon)$ where $\varepsilon > 0$, then I simply point out that $[2 + (\varepsilon/2)]$ is a still smaller number, which is, nonetheless, strictly greater than 2. Thus it is possible to state some minimisation problems, for which there is, in a sense, no solution. Clearly for most practical purposes the answer 2.0000000001 to the above question is satisfactory, but mathematically it is quite simply wrong. More importantly, and ignoring the niceties of mathematical rigour, knowledge of whether there is an exact answer to an optimisation problem is an important factor in determining our approach to it. If we know there is an optimal solution, then we can search for it directly and, more importantly, test any candidate solution for optimality. Suppose, however, we know that for any solution we can always find one better, albeit only marginally better. Then we must adopt a more pragmatic approach, which seeks a good solution upon which any improvement, though possible, is slight. We shall show in the next section that in the case of regular measures optimal schedules do indeed exist.

The meaning of the second question is clear, though perhaps not quite transparent. Naively we might paraphrase it thus. If we have solved a problem for a client and if he then decides the performance measure was inappropriate and so changes it, do we have to do the work all over again? But this is really too simple an interpretation. In section 1.7 I argued—or rather admitted—that we should be over-credulous if we accepted our performance measures as representative of the total costs involved in practical problems; each measure represents only a component of the cost. It would be useful, therefore, to know whether in minimising one component cost we incidentally minimise another. Schedules that minimise several components simultaneously are generally more satisfactory than those that minimise one only. For this reason we investigate the equivalence of performance measures in section 2.3.

2.2 REGULAR MEASURES AND SEMI-ACTIVE SCHEDULES

In Chapter 1 we, or rather I, used the terms 'sequence', 'schedule', and 'timetable' more or less interchangeably. For this section and for part of Chapter 10 it will pay to make some distinction between these terms. We shall say that a processing **sequence** is simply the order in which the jobs are processed through the machines. A processing sequence, therefore, contains no (explicit) information about the times at which the various operations start and finish. Table 1.2 gives an example of a processing sequence. A processing **schedule** does, however, contain timetabling, as well as sequencing information. Thus the Gantt diagram in Fig. 1.1 specifies a complete schedule, each block giving the start and finish times of a particular operation. **Timetabling** is the process whereby we derive a schedule from a sequence.

In showing that an optimal schedule always exists for our problems our approach will be as follows. Firstly we shall note that there are a finite number of distinct processing sequences. Then we shall show that in the case of regular performance measures we need only consider one form of timetabling. Thus, once we agree to use this and only this timetabling, each sequence will define a unique schedule. It follows that we only have a finite number of schedules to consider. Now it is this finiteness that ensures the existence of an optimal schedule. In asking you to find the smallest real number strictly greater than 2, I was asking you to search an infinite set; there are infinitely many numbers slightly greater than 2. Such a search may be unending. But any search of a finite set must eventually end. Here we have a finite set of schedules. So to find one with the smallest value of the performance measure all we need do is search through all the possibilities, comparing one with another until there are no comparisons unmade and the smallest has been found. At least, conceptually that is all we need do. In practice the finite number of schedules may still be too great for such an exhaustive search to be humanly possible. (See the solution to Problem 1.2.2 in section 1.8.) However, it is only the conceptual possibility that we need in order to show that an optimal schedule exists.

So let us begin by noting there are only a finite number of processing sequences. In fact, we have already shown this in our solution to Problem 1.2.2. Still it will do no harm to repeat the argument here. A processing sequence gives for each machine the order of processing jobs. Thus it specifies one permutation of $\{J_1, J_2, \ldots, J_n\}$ for each of the m machines. Since there are $n!$ permutations of $\{J_1, J_2, \ldots, J_n\}$ there are $(n!)^m$ possible processing orders. Actually, because of the technological constraints, many of these may be infeasible. Thus there may be less than $(n!)^m$ processing sequence to consider; but, whatever the case, it is clear that the number is finite.

Timetabling is called **semi-active** if in the resulting schedule there does not exist an operation which could be started earlier without altering the processing sequence or violating the technological constraints and ready dates. In other words, in semi-active timetabling the processing of each operation is started as soon as it can be. If you look back to Figs 1.1, 1.3 and 1.4, you will see that there we used semi-active timetabling without questioning why. It just seemed the sensible thing to do. However, we should question that intuition. Specifically, could Algy and his friends have gone to the country earlier if we had inserted periods of idle time, e.g. if Charles had not picked up the *Guardian* at 10.00 a.m., but had left it for five minutes? Surely not! Yet we must justify this.

Theorem 2.1. In order to minimise a regular measure of performance it is only necessary to consider semi-active timetabling.

Proof. Consider a schedule which has not been constructed by semi-active timetabling. Then there is at least one operation which could be started earlier. Of all such operations choose one with the earliest finishing time. Retimetable this operation to start as early as possible. In this new schedule there cannot be a job whose completion time has increased. Thus the value of a regular measure cannot have increased either.

Repeat this process of retimetabling operations, which could have started earlier, until there are none such left. Note that, because we always pick an operation which has the earliest completion time, no operation can be retimetabled more than once. There are a finite number of operations so this retimetabling process must terminate. The final schedule is the result of semi-active timetabling, since no operation could be started earlier. Moreover, since the value of the performance measure does not increase at any stage, the final schedule is at least as good as the original. Thus for regular measures of performance semi-active timetabling produces schedules at least as good as those we might find by any other method.

N.B. (i) If you find the above proof difficult you may benefit by drawing yourself a Gantt diagram without using semi-active timetabling and then retimetabling operations as described.

N.B. (ii) In the example of Algy and friends we sought to minimise C_{max}, which is a regular measure.

Now the steps of our main argument should be clear. There is only a finite number of feasible processing sequences. For each processing sequence semi-active timetabling clearly produces a unique scheduling. Thus there is only a finite number of these schedules. Theorem 2.1 tells us that in minimising a regular performance measure we need consider no other schedules. Thus we are minimising a function over a finite set and so an optimal schedule exists. There may, of course, be more than one.

Before closing this section, let us briefly return to our distinction between processing sequences and schedules. We shall from now on confine ourselves almost entirely to regular measures and, as a consequence of Theorem 2.1, only consider schedules constructed by semi-active timetabling. It follows that a sequence uniquely defines the associated schedule. Thus there is usually nothing to be gained by pedantically continuing this distinction and we shall not do so.

2.3 RELATIONS BETWEEN PERFORMANCE MEASURES

We shall say that two performance measures are **equivalent** if a schedule which is optimal with respect to one is also optimal with respect to the other and vice versa. In this section we shall prove simple theorems showing that some of the measures introduced in Chapter 1 are equivalent. Let us begin with the simplest.

Theorem 2.2. The following performance measures are equivalent.

$$\text{(i) } \bar{C}, \qquad \text{(ii) } \bar{F}, \qquad \text{(iii) } \bar{W}, \qquad \text{(iv) } \bar{L}.$$

Proof. From the definitions in section 1.5 we have for each job J_i the relations (see especially Fig. 1.2):

$$C_i = F_i + r_i = W_i + \sum_{j=1}^{m} p_{ij} + r_i = L_i + d_i. \qquad (2.1)$$

Summing over the jobs and dividing by n gives the relations between the mean quantities:

$$\bar{C} = \bar{F} + \bar{r} = \bar{W} + (1/n)\sum_{i=1}^{n} \sum_{j=1}^{m} p_{ij} + \bar{r} = \bar{L} + \bar{d} \qquad (2.2)$$

Now the quantities $\bar{r}$, $(1/n \sum_{i=1}^{n}\sum_{j=1}^{m} p_{ij})$, and $\bar{d}$ are constants for each problem and independent of the schedule. Thus in choosing a schedule to minimise $\bar{C}$ we are also minimising $\bar{F}$, $\bar{W}$, and $\bar{L}$. Similarly minimising $\bar{F}$ also minimises $\bar{C}$, $\bar{W}$, and $\bar{L}$, and so on. The four measures are equivalent.

In short, minimising the mean completion time of the jobs also minimises their mean flow time, mean waiting time and mean lateness. We should note here that there is no parallel result concerning C_{max}, F_{max}, W_{max}, and L_{max}; these four measures are not equivalent. In the above proof the step between (2.1) and (2.2) is valid when we take the mean of the quantities involved, but not when we take the maximum. For instance, it is not generally true that $C_{max} = F_{max} + r_{max}$. Consider a 2 job, 1 machine problem with data:

$$J_1\colon r_1 = 0, \quad p_{11} = 5; \qquad J_2\colon r_2 = 10, \quad p_{21} = 1.$$

Fig. 2.1 shows the schedule which processes J_1 then J_2. Here we have

$$C_1 = 5, C_2 = 11 \quad \text{so} \quad C_{max} = 11,$$
$$F_1 = 5, F_2 = 1 \quad \text{so} \quad F_{max} = 5,$$
$$r_1 = 0, r_2 = 10 \quad \text{so} \quad r_{max} = 10;$$

and hence we see $C_{max} = 11 \neq 5 + 10 = F_{max} + r_{max}$. Of course, there are special cases when certain of the measures C_{max}, F_{max}, W_{max}, and L_{max} are equivalent. In problems with all ready times zero C_{max} and F_{max} are equivalent (because $C_i = F_i$ for all jobs). Again, in problems where all the due dates are equal to some constant d, C_{max} and L_{max} are equivalent (because C_i and L_i differ by the same constant d for all jobs).

Let us now note the partial equivalence of L_{max} and T_{max}.

Theorem 2.3. A schedule which is optimal with respect to L_{max} is also optimal with respect to T_{max}.

Proof. $T_{max} = \max\{\max\{L_1, 0\}, \ \max\{L_2, 0\}, \dots, \ \max\{L_n, 0\}\}$
$$= \max\{L_1, L_2, \dots, L_n, 0\}$$
$$= \max\{L_{max}, 0\}.$$

So minimising L_{max} also minimises T_{max}.

N.B. We do *not* claim that minimising T_{max} minimises L_{max}. Any schedule which finishes all the jobs on or before their due dates has

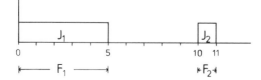

Fig. 2.1—Example showing that $C_{max} \neq F_{max} + r_{max}$.

$T_{max} = 0$, its minimum possible value. However, there may be other schedules which finish the jobs even earlier, so reducing L_{max}, but leaving $T_{max} = 0$.

The next theorem confirms what seems intuitively reasonable. Namely if we minimise the final completion time of all the jobs, then the average number of machines being used at any one time is maximised and the average idle time of a machine is minimised.

Theorem 2.4. The following measures are equivalent:

$$(\text{i}) \ C_{max}, \qquad (\text{ii}) \ \bar{N}_p, \qquad (\text{iii}) \ \bar{I}.$$

(N.B. We seek to minimise C_{max} and $\bar{I}$, but to maximise $\bar{N}_p$.)

Proof. We shall show firstly the equivalence of (i) and (ii), secondly the equivalence of (i) and (iii); the equivalence of (ii) and (iii) then follows immediately.

Equivalence of (i) and (ii):

$$\bar{N}_p = \frac{1}{C_{max}} \int_0^{C_{max}} N_p(t) \, \mathrm{d}t, \quad \text{by definition.}$$

Now $N_p(t)$ is the number of jobs actually being processed at time t.

$$\text{Let} \qquad \delta_i(t) = \begin{cases} 1 & \text{if} \quad J_i \text{ is actually being processed on a} \\ & \qquad \text{machine at time } t; \\ 0 & \text{otherwise.} \end{cases}$$

Then
$$N_p(t) = \sum_{i=1}^n \delta_i(t).$$

Now in the period $[0, C_{max}]$ $\delta_i(t) = 1$ for precisely $\sum_{j=1}^m p_{ij}$, the total processing time of J_i. Thus

$$\int_0^{C_{max}} \delta_i(t) \, \mathrm{d}t = \sum_{j=1}^m p_{ij}.$$

and so
$$\int_0^{C_{max}} N_p(t) \, \mathrm{d}t = \sum_{i=1}^n \int_0^{C_{max}} \delta_i(t) \, \mathrm{d}t = \sum_{i=1}^n \sum_{i=1}^m p_{ij}.$$

Hence
$$\bar{N}_p = \frac{\displaystyle\sum_{i=1}^n \sum_{j=1}^m p_{ij}}{C_{max}}, \qquad (2.3)$$

and it follows that maximising $\bar{N}_p$ is equivalent to minimising C_{max}, since $\sum_{i=1}^n \sum_{j=1}^m p_{ij}$ is a constant independent of the schedule.

Equivalence of (i) and (iii):

Note first that $I_j = C_{max} - \sum_{i=1}^{n} p_{ij}$; that is the idle time on M_j is the difference between the total schedule length and the total processing time on that machine. Thus

$$\bar{I} = \frac{1}{m} \sum_{j=1}^{m} I_j = \frac{1}{m} \left(mC_{max} - \sum_{j=1}^{m} \sum_{i=1}^{n} p_{ij} \right)$$

$$= C_{max} - \frac{1}{m} \sum_{j=1}^{m} \sum_{i=1}^{n} p_{ij}.$$

So C_{max} and $\bar{I}$ differ by a constant independent of the schedule, and hence it follows that minimising C_{max} also minimises $\bar{I}$ and vice versa.

Remembering that C_{max} is a regular measure, we may deduce that the equivalent measure $\bar{I}$ is also regular. Similarly we might conclude that $\bar{N}_p$ is regular and, loosely speaking, this is so. However, to be more precise, we should recall that we seek to minimise a regular measure and so $\bar{N}_p$ cannot be regular, since we wish to maximise it. We may overcome this technicality by the inclusion of a minus sign. Seeking to minimise $(-\bar{N}_p)$ is precisely the same as seeking to maximise $\bar{N}_p$ and now we may correctly deduce from Theorem 2.4 that $(-\bar{N}_p)$ is a regular measure.

Finally we state two theorems concerning $\bar{N}_u$ and $\bar{N}_W$; their proofs are left as Problems 2.5.5 and 2.5.6.

Theorem 2.5. $\bar{N}_u$ and $\bar{C}/C_{max}$ are equivalent measures of performance.

Theorem 2.6. $\bar{N}_W$ and $\bar{W}/C_{max}$ are equivalent measures of performance.

In general $\bar{C}/C_{max}$ and $\bar{W}/C_{max}$ are not regular measures of performance. (See Problem 2.5.7.) Thus we may immediately deduce that neither $\bar{N}_u$ nor $\bar{N}_W$ is regular. However, in the special case of a single machine problem we may deduce regularity. Whatever processing sequence we use C_{max} is a constant, namely the sum of all the processing times. Hence for a single machine problem $\bar{N}_u$ is equivalent to $\bar{C}$ and $\bar{N}_W$ is equivalent to $\bar{W}$. Combining the results with Theorem 2.2 we obtain:

Corollary 2.7. For single machine problems, the following measures are equivalent:

(i) $\bar{C}$,   (ii) $\bar{F}$,   (iii) $\bar{W}$,   (iv) $\bar{L}$,   (v) $\bar{N}_u$,   (vi) $\bar{N}_W$.

2.4 REFERENCES AND FURTHER READING

Conway, Maxwell, and Miller (1967, pp. 9–21, pp. 109–112) and Rinnooy Kan (1976, pp. 16–24) were the principal sources of reference for the above. The concept of a semi-active schedule may be further developed into that of an active schedule. We shall discuss this in Chapter 10. Few authors seem to have considered the equivalence of performance measures, rather they have concentrated their efforts on solving particular problems. Apart from the two books cited above, the only reference of note is that of Gupta and Dudek (1971), where the possible equivalences of certain performance measures and the total scheduling cost are investigated empirically.

2.5 PROBLEMS

1. For weights $\alpha_1, \alpha_2, \ldots, \alpha_n$ show that the following measures are equivalent:

(i) $\sum_{i=1}^{n} \alpha_i C_i$,   (ii) $\sum_{i=1}^{n} \alpha_i F_i$,   (iii) $\sum_{i=1}^{n} \alpha_i W_i$,   (iv) $\sum_{i=1}^{n} \alpha_i L_i$.

2. Define the **mean utilisation** of the machines as the average proportion of the make-span for which the machines are actually processing. Show that maximising mean utilisation is equivalent to minimising make-span.

3. Show that minimising a weighted average idle time, viz, $\sum_{j=1}^{m} \alpha_j I_j$ where $\alpha_j > 0$ and $\sum_{j=1}^{m} \alpha_j = 1$, is equivalent to minimising the unweighted average $\bar{I}$.

4. Consider a 2 job, 1 machine example with processing times 3 and 1 respectively and due dates 3 and 4 respectively. Both are ready for processing at time zero. Evaluate $\bar{L}$ and $\bar{T}$ for the two possible schedules and so deduce that a schedule which is optimal with respct to $\bar{L}$ need not be optimal with respect to $\bar{T}$, nor vice versa.

5. Prove Theorem 2.5.

6. Prove Theorem 2.6.

7. (a) Show $\bar{C}/C_{\max}$ is not a regular measure of performance by considering the $2/2/G/(\bar{C}/C_{\max})$ example with data:

Job	Technological Constraints and Processing Times	
	1st Machine	**2nd Machine**
J_1	$M_2 : p_{12} = 1$	$M_1 : p_{11} = 1$
J_2	$M_1 : p_{21} = 1$	$M_2 : p_{22} = 1$

Both jobs are ready for processing immediately.

Hint: Compare the schedules:

Machine	Processing Sequence		Machine	Processing Sequence	
M_1	J_2	J_1	M_1	J_1	J_2
M_2	J_1	J_2	M_2	J_1	J_2

(b) Show $\bar{W}/C_{max}$ is not a regular measure of performance by considering the $3/3/G/(\bar{W}/C_{max})$ example with data:

	Technological Constraints and Processing Times		
Job	**1st Machine**	**2nd Machine**	**3rd Machine**
J_1	$M_2 : p_{12} = 10$	$M_1 : p_{11} = 10$	$M_3 : p_{13} = 1$
J_2	$M_1 : p_{21} = 1$	$M_2 : p_{22} = 1$	$M_3 : p_{23} = 9$
J_3	$M_3 : p_{33} = 1$	$M_2 : p_{32} = 1$	$M_1 : p_{31} = 1$

The three jobs are ready for processing immediately.
Hint: Compare the schedules:

Machine	Processing Sequence			Machine	Processing Sequence		
M_1	J_2	J_1	J_3	M_1	J_2	J_1	J_3
M_2	J_1	J_2	J_3	M_2	J_1	J_2	J_3
M_3	J_3	J_2	J_1	M_3	J_3	J_1	J_2

Chapter **3**

Single Machine Processing: Basic Results

3.1 INTRODUCTION

We are now in a position to begin solving problems, admittedly not particularly difficult ones as this chapter is only concerned with the very simplest of those involving one machine alone. Nonetheless, we stop classifying, discussing existence and equivalence of solutions and actually attempt to solve something.

We shall assume throughout this chapter that all jobs are ready for processing at the beginning of the processing period. Perhaps it would be best to emphasise both this and the fact that we are only considering single machine problems.

Assumptions for Chapter 3
  (i) $r_i = 0$ for all J_i, $i = 1, 2, \ldots, n$;
  (ii) $m = 1$.

(Generally in each chapter we shall make a number of assumptions over and above those listed in section 1.4. These will apply for the duration of that chapter alone and we shall list them explicitly, as here, in the introduction.)

It is interesting to note that single machine scheduling problems arise in practice more often that one might expect. Firstly, there are obvious ones involving a single machine, e.g. the processing of jobs through a small non-time-sharing computer. Then, there are less obvious ones where a large complex plant acts as if it were one machine, e.g. in paint manufacture the whole plant may have to be devoted to making one colour of paint at a time. Finally, there are job-shops with more than one machine, but in which one machine acts as a 'bottle-neck', e.g. in the hospital example of section 1.3 the treatment of patients may be severely restricted by the shortage of theatre facilities. Thus M_3, the actual surgical operation, may determine the rate of treatment completely, the other hospital facilities easily coping with the demands put upon them. It makes sense to tackle such m-machine problems with single bottle-necks as single machine problems, if only to get a first approximation to their solution.

As a point of notation, since there is only one operation per job, we shall denote by p_i the processing time of job J_i, i.e. we drop the '1' from our earlier notation of p_{i1}.

3.2 PERMUTATION SCHEDULES

In the example of Algy and his friends we saw that certain schedules, in particular the one in Table 1.5, cause the machines, i.e. the papers, to be idle when perhaps they need not be. For instance, Algy could have started to read the *Financial Times* at 8.30, but he leaves it idle so that Charles may have it first. This tactic, not processing the jobs immediately available, but waiting for one that will shortly be so, may be employed by the optimal schedule in the general $n/m/A/B$ problem. However, we shall now see that it is unnecessary to consider such inserted idle time for a single machine problem with a regular measure of performance. In fact, our result follows immediately from Theorem 2.1 (Why?). Nonetheless, there is no harm in proving it afresh from first principles and doing so will emphasise a style of argument that is common to much of our subject.

Theorem 3.1. For an $n/1//B$ problem, where B is a regular measure of performance, there exists an optimal schedule in which there is no inserted idle time, i.e. the machine starts processing at $t = 0$ and continues without rest until $t = C_{\max}$.

Proof. By the argument in section 2.2 we know an optimal schedule exists (B is a regular measure). Let S be an optimal schedule with inserted idle time from $t = \tau_1$ to $t = \tau_2$. Let S' be the schedule obtained from S by bringing forward by an amount $(\tau_2 - \tau_1)$ all operations which start after τ_1. Note that S' is feasible since there is only one machine. Clearly the completion times C_i' under S' are such that

$$C_i' \leq C_i \quad \text{for} \quad i = 1, 2, \ldots, n,$$

where the C_i are the completion times under S. Thus the value of any regular measure cannot increase in passing from S to S'. So S' must be optimal too. This process may be repeated to remove all inserted periods of idle time, so leaving an optimal schedule in which processing is continuous.

In section 1.4 we made the assumption that operations could not be pre-empted; i.e. we are not allowed to interrupt the processing of one operation by the processing of a second before returning to complete that of the first. For the general $n/m/A/B$ problem this is a constraint on our solution; we might do better with a schedule involving pre-emption than we can do with the best one that does not. However, in the single machine case with a regular measure of performance this is not so; there is no advantage to be gained in using pre-emption.

Theorem 3.2. In an $n/1//B$ problem, where B is a regular measure of performance, no improvement may be gained in the optimal schedule by allowing pre-emption.

Proof (Sketch, see Problem 3.7.1). Suppose we have an optimal schedule S in which job I is pre-empted to allow job K to start and then completes without pre-emption at some later time. See the Gantt diagram in Fig. 3.1. We assume K completes without pre-emption. Let S' be the schedule obtained from S by interchanging the first part of I with K. Then no completion times outside the interval $[a,b]$ differ between S and S'. Indeed there is only one change at all in completion time:

$$C'_K < C_K$$

Hence the value of a regular measure cannot increase in passing from S to S'. S is optimal, so S' is also.

By repeating this process the first part of job I can be leap-frogged down the schedule until it is continuous with its last part. Hence there exists an optimal schedule in which job I is not pre-empted. It follows that there exists an optimal schedule entirely without pre-emption.

The significance of Theorems 3.1 and 3.2 is that in $n/1//B$ problems with B regular we need only consider **permutation schedules**. In other words, our task is simply to find a permutation of the jobs $J_1, J_2, \ldots, J_n$ such that, when they are sequenced in that order, the value of B is minimised.

We shall write $J_{i(k)}$ for the job is scheduled at the kth position in the processing sequence and use the subscript $i(k)$ appropriately. Thus J_i is simply a generic job drawn from the list of jobs $\{J_1, J_2, \ldots, J_n\}$ and $J_{i(k)}$ is the job that the processing sequence selects as the kth to be processed, $k = 1, 2, \ldots, n$.

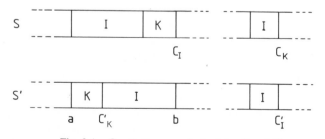

Fig. 3.1—Gantt diagram of schedules S and S'

3.3 SHORTEST PROCESSING TIME SCHEDULING

Suppose that we wish to minimise mean flow time, i.e. our problem is an $n/1//\bar{F}$. For a particular processing sequence:

$$\bar{F} = \frac{1}{n} \sum_{i=1}^{n} F_i = \frac{1}{n} \sum_{i=1}^{n} (W_i + p_i)$$

$$= \frac{1}{n} \sum_{k=1}^{n} (W_{i(k)} + p_{i(k)}) \qquad \text{(on ordering the sum as in the processing sequence)}$$

$$= \frac{1}{n} \sum_{k=1}^{n} W_{i(k)} + \frac{1}{n} \sum_{k=1}^{n} p_{i(k)}.$$

Now $\sum_{k=1}^{n} p_{i(k)} = \sum_{i=1}^{n} p_i$ is a constant for all sequences. Hence to minimise $\bar{F}$ we must minimise $\sum_{k=1}^{n} W_{i(k)}$. If we choose a sequence to make each $W_{i(k)}$ as small as it could possibly be, then we shall clearly minimise their sum.

$W_{i(1)} = 0$ for all sequences, since $J_{i(1)}$ starts immediately.

$W_{i(2)} = p_{i(1)}$, since $J_{i(2)}$ must wait only for $J_{i(1)}$ to be processed. Thus if we choose $J_{i(1)}$ to have the shortest processing time of all the jobs $\{J_1, J_2, \ldots, J_n\}$, we shall minimise $W_{i(2)}$.

$W_{i(3)} = p_{i(1)} + p_{i(2)}$, since $J_{i(3)}$ must wait only for $J_{i(1)}$ and $J_{i(2)}$ to be processed. Thus to minimise $W_{i(3)}$ we choose $J_{i(1)}$ and $J_{i(2)}$ to have the shortest and next shortest processing time from the list $\{J_1, J_2, \ldots, J_n\}$. If we let $J_{i(1)}$ still have the shortest, then we do not affect our earlier minimisation of $W_{i(2)}$. Therefore we can minimise $W_{i(3)}$ and $W_{i(2)}$ simultaneously.

Continuing in this way we build up a schedule in which at the kth job to be processed has the shortest processing time of those remaining. Doing so simultaneously minimises all the waiting times. Thus the result, an **SPT schedule** (Shortest Processing Time), minimises $\bar{F}$. We state this as Theorem 3.3, to which we also provide an alternative proof. The reason for proving this theorem again is that the alternative proof has a form common to many of our later arguments. It does not have the advantage, however, of being as intuitively constructive as the proof above.

Theorem 3.3. For an $n/1//\bar{F}$ problem, the mean flow time is minimised by sequencing such that

$$p_{i(1)} \leq p_{i(2)} \leq p_{i(3)} \leq \ldots \leq p_{i(n)},$$

where $p_{i(k)}$ is the processing time of the job that is processed kth.

Proof (Alternative). Let S be a non-SPT schedule. Then for some k

$$p_{i(k)} > p_{i(k+1)} \qquad (3.1)$$

Let S' be the schedule obtained by interchanging $J_{i(k)}$ and $J_{i(k+1)}$ in S (see

Fig. 3.2). For convenience, label $J_{i(k)}$ as I and $J_{i(k+1)}$ as K. All jobs other than I and K have the same flow times in S' as in S. So the difference in mean flow time for S and S' depends only on the flow times for jobs I and K.

Let
$$a = \sum_{l=1}^{k-1} p_{i(l)} = \begin{cases} W_{\mathrm{I}} \text{ in } S; \\ W'_{\mathrm{K}} \text{ in } S'. \end{cases}$$

So in S, $F_{\mathrm{I}} = a + p_{\mathrm{I}}$ and $F_{\mathrm{K}} = a + p_{\mathrm{I}} + p_{\mathrm{K}}$ (3.2)

Similarly in S', $F'_{\mathrm{I}} = a + p_{\mathrm{K}} + p_{\mathrm{I}}$ and $F'_{\mathrm{K}} = a + p_{\mathrm{K}}$ (3.3)

Hence the contribution of I and K to $\bar{F}$ in S is

$$\frac{1}{n}(F_{\mathrm{I}} + F_{\mathrm{K}}) = \frac{1}{n}(2a + p_{\mathrm{I}} + p_{\mathrm{K}} + p_{\mathrm{I}})$$
$$> \frac{1}{n}(2a + p_{\mathrm{I}} + p_{\mathrm{K}} + p_{\mathrm{K}}) \qquad (p_{\mathrm{I}} > p_{\mathrm{K}} \text{ by } (3.1))$$
$$= \frac{1}{n}(F'_{\mathrm{I}} + F'_{\mathrm{K}}),$$

their contribution to $\bar{F}'$ in S'.

Thus S' has smaller mean flow time than S and so any non-SPT schedule can be bettered. Therefore an SPT schedule must solve an $n/1//\bar{F}$ problem. (N.B. If two or more jobs have equal processing times, there will be more than one SPT schedule.)

From the equivalence of Corollary 2.7 we may also note that SPT scheduling solves the following problems:

$$n/1//\bar{C}, \quad n/1//\bar{W}, \quad n/1//\bar{L}, \quad n/1//N_{\mathrm{u}}, \quad n/1//N_{\mathrm{w}},$$

i.e. it also minimises the mean completion time, the mean waiting time, the mean lateness, the mean number of unfinished jobs, and the mean number of jobs waiting between machines.

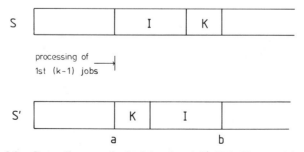

Fig. 3.2—Gantt diagram of schedules S and S'. N.B. Here and in the proof of Theorem 3.3, $J_{i(k)}$ and $J_{i(k+1)}$ have been denoted by I and K respectively.

Example. Given the following $7/1//\bar{F}$ problem:

Job	1	2	3	4	5	6	7
Processing Time	6	4	8	3	2	7	1

we find that the optimal SPT schedule is (7, 5, 4, 2, 1, 6, 3), i.e. do job 7 then job 5, and so on.

To calculate the optimal value of $\bar{F}$ we note that

$$F_{i(1)} = 1$$
$$F_{i(2)} = 1 + 2$$
$$F_{i(3)} = 1 + 2 + 3$$
$$F_{i(4)} = 1 + 2 + 3 + 4$$
$$F_{i(5)} = 1 + 2 + 3 + 4 + 6$$
$$F_{i(6)} = 1 + 2 + 3 + 4 + 6 + 7$$
$$F_{i(7)} = 1 + 2 + 3 + 4 + 6 + 7 + 8$$

$$\frac{1}{7} \sum_{k=1}^{7} F_{i(k)} = \tfrac{1}{7}(7 \times 1 + 6 \times 2 + 5 \times 3 + 4 \times 4 + 3 \times 6 + 2 \times 7 + 8)$$

$$= \tfrac{1}{7} \times 90 = 12\tfrac{6}{7}.$$

Generally we note that for the SPT schedule

$$\bar{F} = \frac{1}{n} \sum_{k=1}^{n} (n - k + 1)p_{i(k)}. \tag{3.4}$$

3.4 EARLIEST DUE DATE SCHEDULING

An initial approach to scheduling is, perhaps, to sequence jobs in the order in which they are required. In other words, to sequence the jobs such that the first processed has earliest due date, the second processed the next earliest due date, and so on. What does this accomplish? The answer is given by:

Theorem 3.4. For an $n/1//L_{max}$ problem, the maximum lateness is minimised by sequencing such that

$$d_{i(1)} \le d_{i(2)} \le d_{i(3)} \le \ldots \le d_{i(n)},$$

where $d_{i(k)}$ is the due date of the job that is processed kth.

Proof. Suppose S is a schedule in which the jobs are not ordered according to increasing due-date. Then for some k, $d_{i(k)} > d_{i(k+1)}$. For convenience, label jobs $J_{i(k)}$ and $J_{i(k+1)}$ as I and K respectively. So we have

$$d_I > d_K \tag{3.5}$$

Let S' be the schedule obtained by interchanging jobs I and K and leaving the rest of the sequence unaltered. (It may help to look back to Fig. 3.2 in the proof of Theorem 3.3.)

Let L be the maximum lateness of the $(n-2)$ jobs other than I and K under S and L' be the corresponding quantity under S'. Clearly $L = L'$. Let L_I, L_K be the lateness of I and K under S and L'_I, L'_K be the corresponding quantities under S'. We, therefore, have the maximum lateness under S

$$L_{max} = \max(L, L_I, L_K),$$

and the maximum lateness under S'

$$L'_{max} = \max(L', L'_I, L'_K)$$
$$= \max(L, L'_I, L'_K), \quad \text{since} \quad L = L'.$$

Now under S

$$L_I = a + p_I - d_I,$$
$$L_K = a + p_I + p_K - d_K, \qquad \text{where} \quad a = \sum_{l=1}^{k-1} p_{i(l)},$$

and under S'

$$L'_K = a + p_K - d_K,$$
$$L'_I = a + p_K + p_I - d_I.$$

Therefore $$L_K > L'_K \quad (\text{since } p_I > 0),$$

and $$L_K > L'_I \quad (\text{by } (3.5)).$$

Hence $$L_K > \max(L'_I, L'_K),$$
$$\Rightarrow \max(L, L_I, L_K) \geq \max(L, L_K)$$
$$\geq \max(L, L'_I, L'_K)$$
$$= \max(L', L'_I, L'_K).$$

So $$L_{max} \geq L'_{max}.$$

Thus any schedule can be rearranged into increasing due date order without increasing its maximum lateness and the theorem is proved.

By Theorem 2.3 we may note that we have also solved the $n/1//T_{max}$ problem. This method of sequencing is called **Earliest Due Date (EDD)**. Often 'earliest' is omitted.

Example. Suppose that we have to solve the $n/1//T_{max}$ problem:

Job	1	2	3	4	5	6
Due Date	7	3	8	12	9	3
Processing Time	1	1	2	4	1	3

Clearly an optimal EDD sequence is $(6, 2, 1, 3, 5, 4)^†$. We may calculate the optimal T_{max} through a tabular form of calculation as below. Here we find $T_{max} = 1$.

Job $J_{i(k)}$	Completion Time $C_{i(k)} = \sum_{l=1}^{k} p_{i(l)}$	Lateness $L_{i(k)} = C_{i(k)} - d_{i(k)}$	Tardiness $T_{i(k)} = \max(0, C_{i(k)})$
6	3	0	0
2	4	1	1
1	5	-2	0
3	7	-1	0
5	8	-1	0
4	12	0	0

N.B. To form the Completion Time column we simply add the processing time of the current job to the completion time of the one preceding it. Thus J_5 has processing time 1, J_3 completed at $t = 7$. So J_5 completes at $t = 8$. This means that instead of $(k - 1)$ additions to form the sum $\sum_{l=1}^{k} p_{i(l)}$ in the kth row of the table, there is only one. We shall use this trivial observation in Chapter 6 when we shall need to calculate $\bar{T}$ for a schedule.

3.5 MOORE'S ALGORITHM

As we have suggested in section 1.5, it occasionally makes sense to penalise tardy jobs equally, however late they are. Thus it would cost just as much to miss a due date by one week as by one hundred years. For instance, Fisher and Jaikumar (1978) use this kind of penalty in scheduling space shuttle flights, since to miss a launch date even by a few hours can completely upset a space mission. If we adopt this philosophy, our scheduling task is simply to minimise the number of tardy jobs; i.e. we face an $n/1//n_T$ problem. We consider an algorithm for solving this due to Moore, but in a form suggested by Hodgson. We first state the algorithm and give an example of its use. Then we shall prove that it does indeed find an optimal schedule.

Algorithm 3.5 (Moore and Hodgson)

Step 1: Sequence the jobs according to the EDD rule to find the *current sequence* $(J_{i(1)}, J_{i(2)}, \ldots, J_{i(n)})$ such that

$$d_{i(k)} \leq d_{i(k+1)} \quad \text{for} \quad k = 1, 2, \ldots, n - 1.$$

$†(2, 6, 1, 3, 5, 4)$ is also optimal.

Step 2: Find the first tardy job, say $J_{i(l)}$, in the current sequence. If no such job is found, go to *Step 4*.

Step 3: Find the job in the sequence $(J_{i(1)}, J_{i(2)}, \ldots, J_{i(l)})$ with the largest processing time and reject this from the current sequence. Return to *Step 2* with a current sequence one shorter than before.

Step 4: Form an optimal schedule by taking the current sequence and appending to it the rejected jobs, which may be sequenced in any order.

N.B. The rejected jobs will all be tardy and these will be the only tardy jobs.

Example. Consider the $6/1//n$ problem:

Job	1	2	3	4	5	6
Due Date	15	6	9	23	20	30
Processing Time	10	3	4	8	10	6

We first form the EDD sequence and compute the completion times until a tardy job is found (Steps 1 and 2):

Current Sequence	2	3	1	5	4	6
Due Date	6	9	15	20	23	30
Processing Time	3	4	10	10	8	6
Completion Time	3	7	17			

Job 1 is the first tardy job in the sequence and of the subsequence $(2, 3, 1)$ it has the largest processing time. So we reject Job 1 (Step 3). We return and repeat Step 2 with the new current sequence:

New Current Sequence	2	3	5	4	6	Rejected Jobs
Due Date	6	9	20	23	30	1
Processing Time	3	4	10	8	6	
Completion Time	3	7	17	25		

Job 4 is the first tardy job in this sequence and of the subsequence $(2, 3, 5, 4)$. Job 5 has the largest processing time. So we reject it (Step 3). Returning to Step 2, we find there are no further tardy jobs:

New Current Sequence	2	3	4	6	Rejected Jobs
Due Date	6	9	23	30	1, 5
Processing Time	3	4	8	6	
Completion Time	3	7	15	21	

Hence we pass to Step 4 and form the optimal sequence (2, 3, 4, 6, 1, 5).
Note that (2, 3, 4, 6, 5, 1) is also optimal. We can lay out the calculation in
a more compact form:

EDD Sequence	2	3	1	5	4	6	Rejected Jobs
Due Date	6	9	15	20	23	30	
Processing Time	3	4	10	10	8	6	
Completion Time	3	7	17				1
Completion Time	3	7	*	17	25		5
Completion Time	3	7	*	*	15	21	

In forming the above table there is one Completion Time row for each
cycle of the algorithm. When a job is rejected, its number is noted in the
right hand column and its completion time in later cycles is simply blanked
out with an asterisk.

We now show that Moore and Hodgson's algorithm does produce an
optimal schedule. First note that any schedule that it produces has the form

$$S_M = (A_M, R_M),$$

where A_M is the sequence of jobs completed on time and ordered by the
          EDD rule,

and R_M is the arbitrary sequence of the jobs that are tardy under S_M.

Our first step is to show that any optimal schedule S can be taken to have a
similar form $S = (A,R)$, although we do not claim that the jobs in A are
necessarily the same as those in A_M.

Let S be any optimal schedule and R the set of jobs tardy under S.
Resequence S so that the jobs in R are scheduled last and the order of the
remaining jobs is unchanged. Let $S' = (B,R)$ be the sequence so formed.
Note that the jobs in B are started no later in S' than they were in S. Thus
no job in B can be tardy and the number of jobs tardy under S' is no
greater than under S. Next observe that, since there are no tardy jobs in B,
$L_{max} \leq 0$ for this subsequence. Since the EDD rule minimises L_{max}
(Theorem 3.4), resequencing B according to EDD does not increase L_{max}.
Thus, if A is the result of resequencing B in this way, no job scheduled
according to A can have positive lateness, i.e. can be tardy. So the number
of jobs tardy under $S'' = (A,R)$ is no greater than under S. S is optimal;
thus S'' is also. Hence we have an optimal schedule of the required form.

The next stage of our proof is to show that the number of late jobs
under a schedule produced by Moore and Hodgson's algorithm is no

greater than that under any optimal schedule. So let

$$S_M = (A_M, R_M) \text{ be a schedule produced by Moore's algorithm,}$$
$$S = (A, R) \quad \text{be an optimal schedule,}$$
and $\quad\quad\quad v \quad\quad\quad$ be the number of jobs late under S_M.

It will help our notation to renumber all the jobs so that $\{J_1, J_2, \ldots, J_n\}$ is in EDD order, i.e. so that

$$d_i \le d_{i+1} \quad \text{for} \quad i = 1, 2, \ldots, (n-1).$$

Thus $(J_1, J_2, \ldots, J_n)$ is the sequence of jobs produced by Step 1 of the algorithm. Let $J_{l(1)}, J_{l(2)}, J_{l(3)}, \ldots, J_{l(v)}$ be the sequence of 'next late jobs' found in Step 2. Let $J_{r(1)}, J_{r(2)}, \ldots, J_{r(v)}$ be the sequence of the jobs rejected by Step 3. Thus $J_{r(i)}$ is the job that is rejected in the ith cycle on discovering that $J_{l(i)}$ is the next late job in the current sequence. To show S_M is optimal we shall compare the scheduling of jobs $\{J_1, J_2, J_3, \ldots, J_{l(i)}\}$ for $i = 1, 2, \ldots, v$ under S_M and S. We need one more piece of notation. Let μ_i be the number of jobs in $\{J_1, J_2, \ldots, J_{l(i)}\}$ that are tardy under S. We shall show that $\mu_i \ge i$ for $i = 1, 2, \ldots$ and hence that the number of jobs tardy under S_M is no greater than that under S.

Consider first the case for $i = 1$. By construction of $J_{l(1)}$, it is impossible to schedule the jobs $\{J_1, J_2, J_3, \ldots, J_{l(1)}\}$ without at least one being tardy. Thus

$$\mu_1 \ge 1.$$

Also note that the job rejected by S_M, namely $J_{r(1)}$, has the longest processing time of the sequence $\{J_1, J_2, \ldots, J_{l(1)}\}$. Pair this job with the job, say $J_{q(1)}$, of the μ_1 jobs rejected by S which has the longest processing time. Thus it is immediate that $p_{q(1)} \le p_{r(1)}$, i.e. the job rejected by S_M has a processing time at least as long as the longest of the μ_1 rejected by S.

Next assume that for some $i \ge 1$, we have shown that:

(A) $\mu_i \ge i$;
(B) $J_{r(1)}, J_{r(2)}, \ldots, J_{r(i)}$ can be paired with jobs $J_{q(1)}, J_{q(2)}, \ldots, J_{q(i)}$ in the μ_i rejected by S such that

$$p_{r(k)} \ge p_{q(k)} \quad \text{for} \quad k = 1, 2, \ldots, i.$$

We shall show (A) and (B) also hold for $(i + 1)$. Since we have shown they are true for $i = 1$, we can now claim these relations together to show by induction that A and B hold for $i = v$.

It is trivial that $\mu_{i+1} \ge \mu_i$. So, if $\mu_i > i$ we have $\mu_{i+1} \ge \mu_i \ge i + 1$, since μ_i is integral. Thus assume $\mu_i = i$. By our pairing we have that the jobs rejected under S_M and S are such that

$$\sum_{k=1}^{i} p_{r(k)} \ge \sum_{k=1}^{i} p_{q(k)} = \sum_{k=1}^{\mu_i} p_{q(k)}.$$

Thus the 'time saved' by the rejections under S_M is at least as great as that under S. Hence under S_M jobs ($J_{l(i+1)}$, $J_{l(i+2)}$, ..., $J_{l(i+1)}$) start no later than they do under S. By construction of $J_{l(i+1)}$ we know one of these jobs must be tardy under S_M, so at least one must also be late under S. Thus at least one of these jobs must be rejected by S and μ_{i+1} must therefore be at least one greater than μ_i. So

$$\mu_{i+1} \geq \mu_i + 1 \geq i + 1 \quad \text{as required.}$$

Given (A) and (B) we have shown $\mu_{i+1} \geq i + 1$, i.e. (A) holds for $i + 1$. We must now show that the pairing condition in (B) also holds for $i + 1$. To do this we simply pair the job $J_{r(i+1)}$ with the longest unpaired job $J_{q(i+1)}$ of the μ_{i+1} jobs rejected by S. Because $J_{r(i+1)}$ is the longest job so far unrejected by S_M in the sequence $J_1, J_2, \ldots, J_{l(i+1)}$, and because $J_{r(1)}, J_{r(2)}, \ldots, J_{r(i)}$ have all been paired with the longest then unpaired jobs amongst those rejected by S, we see that $p_{r(i+1)} \geq p_{q(i+1)}$ as required.

We have now all but finished. We have shown:

$$\mu_1 \geq 1,$$
$$\Rightarrow \quad \mu_2 \geq 2,$$
$$\Rightarrow, \quad \text{in turn,} \quad \mu_3 \geq 3,$$
$$\vdots$$
$$\Rightarrow, \quad \text{finally,} \quad \mu_v \geq v.$$

Thus S rejects at least as many jobs as S_M. S is optimal; therefore S_M is also. We have proved:

Theorem 3.6. Algorithm 3.5 finds an optimal schedule for an $n/1//n_T$ problem.

3.6 REFERENCES AND FURTHER READING

Conway, Maxwell and Miller (1967) and Baker (1974) deal with single machine processing in considerable detail and indeed we shall discuss the topic further ourselves, principally in Chapters 4 and 6. Rinnooy Kan (1976) and Graham et al. (1979) survey the majority of the results in the area.

The SPT sequencing result is very old and has been proved in a number of ways: Conway, Maxwell and Miller (1967, p. 26) give an interesting and instructive proof based upon the area under a concave graph, and Rinnooy Kan (1976, p. 68) provides an algebraic proof quite distinct from either of ours above. The EDD sequencing result was originally due to Jackson (1955). The development of Moore and Hodgson's algorithm as originally given by Moore (1968) was much longer than that given above. Ours is based upon one of Sturm's (1970). Kise et al. (1978) have modified and extended this algorithm so that it solves $n/1//n_T$ problems in which there are non-zero ready times, but with the restriction that $d_i \leq d_j \Rightarrow r_i \leq r_j$ (i.e. the ready times must increase in the same sequence as the due dates). Sidney (1973) has modified Moore's result in a different direction. His algorithm allows that certain jobs be specified as non-tardy, i.e. the processing sequence must ensure that these are completed before their due dates.

3.7 PROBLEMS

1. Complete and proof of Theorem 3.2 by:
(a) allowing that job I might be broken into more than two segments by pre-emption; and
(b) allowing that the job K may itself be pre-empted.
2. Use the method of the alternative proof of Theorem 3.3 to show SPT scheduling solves the $n/1//(1/n \ \Sigma_{i=1}^{n} F_i^2)$ problem, i.e. it also minimises mean square flow time.
3. Prove formula (3.4) above and find a similar expression for $\bar{W}$.
4. Solve the $n/1//\bar{C}$ problem below and for the optimal schedule find:
(i) $\bar{C}$, (ii) $\bar{N}_u$.

Job	1	2	3	4	5	6	7	8
Processing Time	10	2	4	7	3	1	3	2

5. You are asked to solve the problem $n/1//\Sigma \alpha_i F_i$ where $\Sigma_{i=1}^{n} \alpha_i = 1$ and $\alpha_i > 0$ for $i = 1, 2, \ldots, n$, i.e. you have to minimise a weighted mean of the flow times. Show that an optimal schedule sequences the jobs such that

$$\frac{p_{i(1)}}{\alpha_{i(1)}} \leq \frac{p_{i(2)}}{\alpha_{i(2)}} \leq \frac{p_{i(3)}}{\alpha_{i(3)}} \leq \ldots \leq \frac{p_{i(n)}}{\alpha_{i(n)}}$$

What would the optimal sequence be if some of the $\alpha_i = 0$?
6. Solve the $n/1//L_{max}$ problem and find the optimal L_{max}.

Job	1	2	3	4	5	6	7
Due Date	80	20	67	48	100	30	50
Processing Time	29	13	31	20	7	3	9

7. At *any* time t define the **slack time** of an unprocessed job J_i as $(d_i - p_i - t)$, i.e. the amount of time remaining before J_i must be started in order that it is finished on time. Consider sequencing according to non-decreasing slack time. In other words, when one job completes select the next to be the one which has the shortest slack time. Show that this sequence *maximises* the *minimum* job lateness.

8. Solve the $10/1//n_T$ problem:

Job	1	2	3	4	5	6	7	8	9	10
Due Date	19	16	25	3	8	14	31	23	2	15
Processing Time	5	3	1	2	4	4	2	1	1	4

9. N.A.S.A. has one space shuttle with which to launch 8 space stations. Each station is specifically designed to perform certain astronomical observations. Each must be placed in position by a certain date or it will be useless. Given the data below, in what order should the stations be launched? Assume that the launching and construction sequence starts from 1st January 1982.

Station	Time to load into shuttle, launch, and build in space	Must be in orbit by 1st
1	1 year, 2 months	Apr. 1986
2	7 months	Jan. 1983
3	11 months	Aug. 1983
4	3 months	Mar. 1986
5	1 year, 8 months	Sept. 1985
6	4 months	Aug. 1982
7	7 months	Dec. 1982
8	1 year, 2 months	June 1984

10. Consider an $n/1//L_{max}$ problem in which the ready times are not necessarily zero but must be integral and in which the processing times are unity (i.e. all $p_i = 1$). Construct an algorithm for finding an optimal schedule.

Hint: Note that, since r_i are integral and the $p_i = 1$, no job can become ready during the processing of another.

Chapter **4**

Single Machine Processing: Precedence Constraints and Efficiency

4.1 INTRODUCTION

Few firms consider all their customers to be of equal importance. For example, suppose that there is a number of jobs to be processed and that one is for a customer of exceptional importance, whose goodwill must be maintained at all costs. In this case, the firm is extremely likely to decide that his job should be rushed through and finished first, whatever the consequences for the scheduling of the other jobs. Here we have an example of a scheduling problem which is complicated by the existence of **precedence constraints**. In general, these limit the choice of schedule by demanding that certain subsets of jobs be processed in a given order. There are many other practical scheduling problems in which precedence constraints arise, but it will be sufficient for our purposes to illustrate just two further here. First, consider a situation where any adjustment to a machine's settings may require a substantial time to take effect; perhaps certain components need to warm up or cool down. In such a case it makes sense to process together any group of jobs that require similar settings, because the change-over times between the jobs will be slight. Second, consider the scheduling of programs upon a computer. Suppose that one program produces an output file that a second requires as input data. Then obviously the schedule must ensure that the first is fully executed before the second is begun. In both examples there are clear precedence constraints upon the scheduling of the jobs.

However natural it may at first seem to model certain aspects of a scheduling problem using precedence constraints, it is often better theoretically to use other devices. Ideally, important customers and similar complications should be treated by an appropriate choice of performance measure, one that penalises a schedule very heavily if a particular job is tardy. Similarly, when the set-up time of an operation depends crucially upon which job was processed immediately before, we should not introduce precedence constraints into our model. Rather we should abandon

Assumption 5 of section 1.4 and admit instead that the processing times, which include the set up times, are not necessarily sequence-independent. However, the correctly formulated problems may be very difficult to solve, whereas the presence of precedence constraints may actually help in the solution of a problem because they reduce the number of feasible sequences. But this does not mean that you should think of precedence constraints as solely a trick by which we may avoid particularly difficult problems. In very many cases they enter a problem quite naturally. For instance, in the computing example it is not because of cost that we seek to run the second program after the first; it is simply because of a logical necessity.

It is understandable that some people should confuse technological constraints and precedence constraints; so we shall pause and try to make their distinction clear. Technological constraints give the order in which the operations that compose each job must be processed. In short, they give the routing that each job must follow between the machines. Precedence constraints, on the other hand, restrict the sequence of processing operations between different jobs. They insist that a certain operation of one job must be fully processed before a certain operation of a different job may be begun. Often, but by no means always, the restriction applies to the last operation of the former job and the first operation of the latter, so insisting that the one job is completed before the other is begun. Perhaps the easiest way to appreciate the distinction between technological and precedence constraints is to consider single machine processing. Because there is only one operation per job, there can be no technological constraints here. However, there can quite clearly be precedence constraints; each of the three examples quoted above can apply to the single machine case.

In this chapter we consider two simple classes of single machine problems with precedence constraints. In Section 6.4 we consider another class of such problems. Finally, in Chapter 12 we further discuss the form that precedence constraints may take and extensively reference the literature for methods of treating them in the flow-shop and job-shop, as well as in more general scheduling problems.

Later in this chapter we turn from precedence constraints to **efficiency**. This topic can best be introduced by referring back to our discussion in Section 1.7. There we admitted that our choice of performance measure seldom represented the total cost in a scheduling problem, but only a component. For instance, if we seek to minimise the maximum tardiness of a schedule, then we will be reducing in a general sense the penalty costs incurred by the late completion of jobs, but ignoring any inventory or utilisation costs that might also be incurred, the idea of efficiency takes us a small, but significant step from the minimisation of a single component cost towards the minimisation of the total cost.

Suppose that in a single machine problem we feel that T_{max} is a suitable indicator of the penalty costs arising from late completion of jobs and that $\bar{F}$ is an equally suitable indicator of in-process inventory costs (see Corollary 2.7). Suppose further that we feel that any other costs can be safely neglected. Thus we are saying that the total cost is a function of T_{max} and $\bar{F}$ alone, say $c(T_{max}, \bar{F})$. Now it is reasonable to suppose that if either T_{max} or $\bar{F}$ increases, then the total cost does too. So we may assume that $c(T_{max}, \bar{F})$ is an increasing function of both its arguments. Suppose we have to choose between two schedules S and S' for which $T_{max} < T'_{max}$ and $\bar{F} < \bar{F}'$, where in an obvious notation T'_{max} and $\bar{F}'$ refer to S'. Then it is clear that we should prefer S because we have $c(T_{max}, \bar{F}) < c(T'_{max}, \bar{F}')$ (by the increasing nature of c). Indeed, a little thought shows that we can always say that S is better then S', or S **dominates** S', if

$$\left. \begin{array}{c} T_{max} \le T'_{max} \\ \bar{F} \le \bar{F}' \end{array} \right\} \begin{array}{l} \text{with strict inequality holding} \\ \text{in at least one case} \end{array} \qquad (4.1)$$

We shall say that a schedule S' is **efficient** if there does not exist a schedule S which dominates it, i.e. such that (4.1) holds.

To be strictly correct, when saying a schedule is efficient, we should also state the performance measures concerned. In general, a schedule S' is **efficient with respect to measures $\mu_1, \mu_2, \ldots, \mu_\nu$,** if there does not exist a schedule S such that:

$$\left. \begin{array}{c} \mu_1 \le \mu'_1 \\ \mu_2 \le \mu'_2 \\ . \\ \vdots \\ \mu_\nu \le \mu'_\nu \end{array} \right\} \begin{array}{l} \text{with strict inequality holding} \\ \text{in at least one case.} \end{array} \qquad (4.2)$$

N.B. It is rather pointless to consider efficiency with respect to a set of equivalent measures. (*Why?*) We assume, therefore, that no pair of the measures $\mu_1, \mu_2, \ldots, \mu_\nu$ are equivalent.

If we can identify the set of efficient schedules with respect to performance measures $\mu_1, \mu_2, \ldots, \mu_\nu$, then we may have made a useful step towards finding the schedule which minimises total cost. For, if the total cost is an increasing function of $\mu_1, \mu_2, \ldots, \mu_\nu$ alone, the minimum cost schedule must be efficient. In Section 4.5 we return to our example above and for a single machine problem find the set of schedules that are efficient with respect to T_{max} and $\bar{F}$. Having found this efficient set, we determine an optimal schedule for a particular linear total cost function.

As indicated above, we shall study both this chapter's topics within the context of single machine problems. Moreover, to maintain simplicity keep the analysis simple we assume that all the jobs are ready for processing immediately. Thus we continue to make the assumptions introduced in Chapter 3:

Assumptions for Chapter 4
(i) $r_i = 0$ for all J_i, $i = 1, 2, \ldots, n$;
(ii) $m = 1$.

4.2 REQUIRED STRINGS OF JOBS

In the following we shall assume that the n jobs have been partitioned into K **strings** of $n_1, n_2, \ldots, n_K$ jobs respectively. Moreover, we shall assume that the processing order of jobs within a string is given and that once a string of jobs is started then it must be completed before jobs in another string are begun. Such required strings of jobs are also called **chains** of jobs by some authors. We shall use a notation peculiar to this section.

Let p_{ij} be the processing time of the jth job in the ith string; $p_i' = \Sigma_{j=1}^{n_i} p_{ij}$, be the total processing time of the ith string; and F_{ij} be the flow time of the jth job in the ith string. Note that F_{in_i} is the flow time of the ith string.

We shall assume that our task is to minimise the mean job flow time, i.e. to solve $n/1//\bar{F}$ under the precedence constraints given by the required strings. Note that, if our task was to minimise the mean *string* flow time (viz. $1/K \ \Sigma_{i=1}^{K} F_{in_i}$), we would sequence the strings according to the SPT rule applied to $p_1', p_2', \ldots, p_K'$. For then we could consider the strings as composite jobs. However our performance measure is the mean *job* flow time (viz. $1/n \ \Sigma_{i=1}^{K} \ \Sigma_{j=1}^{n_i} F_{ij}$).

Theorem 4.1. For the $n/1//\bar{F}$ problem with required strings of jobs whose processing order is fixed in advance and whose processing cannot be preempted once started, the optimal schedule(s) are such that

$$\frac{p_{i(1)}'}{n_{i(1)}} \leq \frac{p_{i(2)}'}{n_{i(2)}} \leq \frac{p_{i(3)}'}{n_{i(3)}} \leq \ldots \leq \frac{p_{i(K)}'}{n_{i(K)}} \tag{4.3}$$

where the $i(1)$th string is scheduled first, and $i(2)$th string is scheduled second, etc.

Proof. Suppose that S is a schedule such that (4.3) does not hold. Then there are two strings Q and R adjacent in S such that

$$p_Q'/n_Q > p_R'/n_R \tag{4.4}$$

Let S' be the schedule formed by interchanging strings Q and R (see Fig. 4.1). The resulting change in the total flow time arises from two sources:
(i) the contribution from string Q increases by the amount

$$n_Q p_R',$$

since each job in Q has its flow time increased by p_R';
(ii) the contribution from string R decreases by the amount

$$n_R p_Q',$$

since each job in R has its flow time decreased by p_Q'.

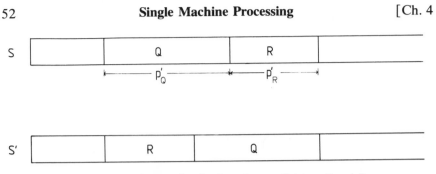

Fig. 4.1 Gantt diagram showing interchange of strings Q and R.

There is no change to the flow time of any other job. From (4.4) we have $n_R p'_Q > n_Q p'_R$. Therefore the interchange of strings Q and R brings a net decrease in the total flow time and, hence, in the mean flow time.

It is a simple matter to apply this theorem in a problem to deduce an optimal schedule. The only remark possibly worth making is that a job which does not lie in any string should be treated as a degenerate string with $n_i = 1$.

4.3 LAWLER'S ALGORITHM

In this section we discuss an algorithm which was developed by Lawler (1973) to deal with rather more general precedence constraints. Here we shall simply be constrained to process certain jobs before, but not necessarily immediately before, certain others. Lawler's algorithm minimises the maximum cost of processing a job, where this cost has a general form $\gamma_i(C_i)$ for J_i and is taken to be non-decreasing in the completion time C_i. Thus the algorithm minimises

$$\max_{i=1}^{n}\{\gamma_i(C_i)\}. \tag{4.5}$$

Because of the non-decreasing nature of each $\gamma_i(C_i)$, it follows immediately that this performance measure is regular. At first sight (4.5) looks somewhat foreboding compared with our earlier performance measures; however, for specific choices of $\gamma_i(C_i)$ (4.5) takes on much more familiar forms. If $\gamma_i(C_i) = C_i - d_i = L_i$, then (4.5) gives the measure L_{max}; if $\gamma_i(C_i) = \max\{C_i - d_i, 0\}$, it gives T_{max}. To develop the algorithm we need the following theorem.

Theorem 4.2. Consider the $n/1//\max_{i=1}^{n}\{\gamma_i(C_i)\}$ problem with precedence constraints. Let V denote the subset of jobs which may be performed last, i.e. those jobs which are not required to precede any other. Note that the final job in the schedule must complete at $\tau = \Sigma_{i=1}^{n} p_i$. Let J_k be a job in V

such that

$$\gamma_k(\tau) = \min_{J_i \text{ in } V}\{\gamma_i(\tau)\},$$

i.e. of all the jobs that may be performed last J_k incurs the least cost. Then there is an optimal schedule in which J_k is scheduled last.

Proof. Suppose S is an optimal schedule with J_k not last. We assume that S is compatible with the precedence constraints. Then S has the form

$$(A, J_k, B, J_l),$$

where J_l is the last job under S and A and B are subsequences of the other $(n-2)$ jobs. (N.B. either A or B could be empty). Consider the new sequence:

$$S' = (A, B, J_l, J_k).$$

Since S obeyed the precedence constraints and J_k may be last, S' is feasible. Remember that $\gamma_i(C_i)$ is non-decreasing in C_i. Since all completion times of jobs other than J_k have decreased in passing from S to S', no cost other than for J_k can have increased. By construction J_k is chosen such that

$$\gamma_k(\tau) = \min_{J_i \text{ in } V}\{\gamma_i(\tau)\}$$
$$\leq \gamma_l(\tau).$$

Therefore, even if the cost of J_k increases under S', it does not exceed the cost of J_l under S. Thus the maximum cost under S' is no greater than under S. Hence we may choose an optimal schedule of the form required.

It is now a relatively simple matter to deduce the form of the algorithm. Let J_k be the job which by Theorem 4.2 may be last in an optimal sequence. Thus there is an optimal sequence of the form (A, J_k), where A is a permutation of the other $(n-1)$ jobs. The maximum cost of this sequence (A, J_k) is the larger of $\gamma_k(\tau)$, the cost of completing J_k last, and the maximum cost of completing the jobs in A. An optimal sequence can be found by making both these terms as small as possible. J_k has been chosen so that $\gamma_k(\tau)$ is the minimum for all the jobs that could be processed last. So to construct an optimal sequence our task is simply to choose A so that the maximum cost of completing its jobs is as small as possible. In other words, we face the problem of scheduling $(n-1)$ jobs subject to precedence constraints so that the maximum cost of the individual jobs is minimised. We face a new problem with the same form as our original, but with $(n-1)$ jobs instead of n. By Theorem 4.2 we can say which job should be last for this new problem and, hence, $(n-1)$st in the sequence for the original problem. We are left with the task now of scheduling $(n-2)$ jobs. And so we go on, repeatedly scheduling a job at the end of the sequence and reducing the size of the problem by 1. Eventually we completely solve the original problem.

Example. Consider the $6/1//L_{max}$ problem with data below and the constraints that J_1 must precede J_2 which must in turn precede J_3, and also that J_4 must precede both J_5 and J_6. We show those constraints graphically in Fig. 4.2.

Fig. 4.2 Precedence constraints for the example.

Job	J_1	J_2	J_3	J_4	J_5	J_6
Processing Time	2	3	4	3	2	1
Due Date	3	6	9	7	11	7

Finding the job processed 6th: $\tau = 2 + 3 + 4 + 3 + 2 + 1 = 15$. Jobs Jobs J_3, J_5 and J_6 can be processed last, i.e. $V = \{J_3, J_5, J_6\}$. So the minimum lateness over $V = \min\{(15-9), (15-11), (15-7)\}$, which occurs for J_5. Hence J_5 is scheduled 6th.

Finding the job processed 5th: We delete J_5 from our list and note that the completion time of the first five jobs is $\tau = 15 - 2 = 13$. Jobs J_3 or J_6 can be processed last now; i.e. $V = \{J_3, J_6\}$. So the minimum lateness over $V = \min\{(13 - 9), (13 - 7)\}$, which occurs for J_3. So job J_3 is scheduled 5th.

Finding the job processed 4th: We have now deleted J_3 and J_5 from our list. Thus J_2 becomes available for processing last. We have $\tau = 13 - 4 = 9$ and $V = \{J_2, J_6\}$. Minimum lateness over $V = \min\{(9 - 6), (9 - 7)\}$, which occurs for J_6. Thus J_6 is scheduled 4th.

Finding the job processed 3rd: J_3, J_5 and J_6 have now been deleted; hence we now have $V = \{J_2, J_4\}$ and $\tau = 9 - 1 = 8$. Minimum lateness over $V = \min\{(8 - 6), (8 - 7)\}$, which occurs for J_4. thus J_4 is scheduled 3rd.

The jobs scheduled first and second are now automatically J_1 and J_2 respectively, because of the precedence constraints. These calculations may be laid out concisely as in the following table.

τ	J_1	J_2	J_3	J_4	J_5	J_6	**Scheduled Job**
15	*	*	6	*	④	8	J_5
13	*	*	④	*	S	6	J_3
9	*	3	S	*	S	②	J_6
8	*	2	S	①	S	S	J_4
5	*	⊘−1	S	S	S	S	J_2
2	⊘−1	S	S	S	S	S	J_1

The entries in column J_i are respectively:

*   if J_i cannot be scheduled last,

or

$\tau - d_i$ if it is possible to schedule J_i last,

or

S   it J_i has already been scheduled.

The scheduled job at each τ has the minimum lateness $(\tau - d_i)$ and this value is circled in the table. The final schedule may be found by reading *up* the final column and the value of L_{max} found by taking the largest of the circled quantities. Here we find the schedule (J_1, J_2, J_4, J_6, J_3, J_5) with $L_{max} = 4$.

We shall leave our study of precedence constraints for a while; they are discussed again in Chapter 6. Now we turn to the topic of efficiency.

4.4 SPT SEQUENCING SUBJECT TO DUE-DATE CONSTRAINTS

The problem that we consider in this and the next section is how to find schedules that are efficient with respect to $\bar{F}$ and T_{max}. In this section we develop an algorithm due to Smith (1956) which, although not directed to the immediate solution of the problem, is nonetheless instrumental in the construction of efficient schedules. Thus we shall not explicitly mention efficiency again until the next section; rather we follow Smith in the motivation of his theory.

Suppose that we have a $n/1//T_{max}$ problem to solve. By Theorem 3.4 we know that a solution may be found by the EDD rule. Suppose that when we construct this schedule we find that $T_{max} = 0$, i.e. that all the due dates can be met. Then, before we actually process the jobs in EDD sequence, perhaps we should stop and think. There might be schedules other than the

EDD which also satisfy the due-date constraints. Might we prefer one of these? For instance, one might give a smaller value of $\bar{F}$ than the EDD sequence. Indeed this is more than possible; later we shall give an example to show this. Smith's algorithm gives us a way of finding a schedule to minimise $\bar{F}$ subject to the condition that $T_{max} = 0$.

As does Lawler's, Smith's algorithm builds up a schedule from back to front; it first finds a job to be nth in the processing sequence, then $(n - 1)$st, and so on. Again as with Lawler's, it is based upon a theorem which describes the characteristics of a job that may be processed last.

Theorem 4.3. For the n job, 1 machine problem such that all the due dates can be met, there exists a schedule which minimises $\bar{F}$ subject to $T_{max} = 0$ and in which job J_k is last, if and only if

(i) $d_k \geq \sum_{i=1}^{n} p_i$,

and (ii) $p_k \geq p_l$ for all jobs J_l such that $d_l \geq \sum_{i=1}^{n} p_i$.

Proof. Consider a schedule S in which job J_k is last and conditions (i) and (ii) hold. Suppose we interchange J_k and some job J_l in the sequence. If $d_l < \sum_{i=1}^{n} p_i$, i.e. if condition (i) does not hold for job J_l, then J_l will be tardy and so T_{max} will no longer be zero. On the other hand, if $d_l \geq \sum_{i=1}^{n} p_i$ but $p_k > p_l$, i.e. if condition (ii) does not hold for the *new* sequence, then we shall show that the mean flow time of the sequence is increased. Let the schedule with J_k and J_l interchanged be S'. Then, using a prime to indicate quantities connected with S', we have:

$$F'_k > F_l, \quad \text{since} \quad p_k > p_l;$$

$$F'_l = F_k = \sum_{i=1}^{n} p_i;$$

and, moreover, the flow time of any job scheduled between J_k and J_l will be increased by $(p_k - p_l) > 0$. The flow times of all earlier jobs will be unchanged. It follows immediately that $\bar{F}' > \bar{F}$. Thus the theorem is proved.

It is now a straight-forward matter to develop Smith's Algorithm. We omit the arguments underlying its development, since they are entirely parallel to those we used when constructing Lawler's Algorithm. Thus we simply state:

Algorithm 4.4 (Smith)

Step 1: Set $k = n$, $\tau = \sum_{i=1}^{n} p_i$; $U = \{J_1, J_2, J_3, \ldots, J_n\}$.
Step 2: Find $J_{i(k)}$ in U such that (i) $d_{i(k)} \geq \tau$ and (ii) $p_{i(k)} \geq p_l$ for all J_l in U such that $d_l \geq \tau$.
Step 3: Decrease k by 1; decrease τ by $p_{i(k)}$; delete $J_{i(k)}$ from U.

Step 4: If there are more jobs to schedule, i.e. if $k \geq 1$, go to *Step 2*. Otherwise stop with the optimal processing sequence $(J_{i(1)}, J_{i(2)}, \ldots, J_{i(n)})$.

In stating the algorithm we have used the following notation:

k is the position in the processing sequence currently being filled (N.B. k cycles *down* n, $(n - 1)$, ..., 1);

τ is the time at which the job kth in the sequence must complete;

and U is the set of unscheduled jobs.

There are two remarks that should be made upon the working of this algorithm. First, it is implicit in our development that, before using the algorithm, we check the EDD sequence to see whether there is a schedule with $T_{max} = 0$. In fact, we need not do this. If there is no schedule with $T_{max} = 0$, we shall discover this in working through Smith's algorithm. At some pass through Step 2 we shall be unable to find any $J_{i(k)}$ with $d_{i(k)} \leq \tau$. Second, sometimes $J_{i(k)}$ may not be determined uniquely; there may be two or more jobs for which conditions (i) and (ii) are true. If this is so, we make the choice of $J_{i(k)}$ arbitrarily—at least for the present.

Example. Consider solving the $4/1//\bar{F}$ problem subject to $T_{max} = 0$ when the data is as below.

Job	J_1	J_2	J_3	J_4
Processing Time	2	3	1	2
Due Date	5	6	7	8

Applying the algorithm we take the following steps:

Step 1: $k = 4$, $\tau = 8$, $U = \{J_1, J_2, J_3, J_4\}$.
Step 2: Only J_4 satisfies condition (i) so we choose $J_{i(4)} = J_4$.
Step 3: $k = 3$; $\tau = 6$; $U = \{J_1, J_2, J_3\}$.
Step 4: $k \geq 1$.
Step 2: J_2 and J_3 satisfy condition (i); J_2 has the larger processing time, so $J_{i(3)} = J_2$.
Step 3: $k = 2$; $\tau = 3$; $U = \{J_1, J_3\}$.
Step 4: $k \geq 1$.
Step 2: J_1 and J_3 satisfy condition (i); J_1 has the larger processing time, so $J_{i(2)} = J_1$.
Step 3: $k = 1$; $\tau = 1$; $U = \{J_3\}$.
Step 4: $k \geq 1$.
Step 2: J_3 satisfies condition (i) so $J_{i(1)} = J_3$.
Step 3: $k = 0$; $\tau = 0$; U is empty.
Step 4: Optimal sequence in (J_3, J_1, J_2, J_4).

N.B. the sequence has $\bar{F} = 18/4$, whereas the EDD sequence (J_1, J_2, J_3, J_4)

has $\bar{F} = 21/4$. Thus we have the example promised earlier which shows that the EDD sequence need not minimise $\bar{F}$ subject to $T_{max} = 0$.

Smith left his theory here, but Van Wessenhove and Gelders (1980) have developed his ideas further. The rationale underlying Smith's approach is that we are only willing to consider reducing $\bar{F}$ once we have ensured that $T_{max} = 0$. In other words, penalty costs have overriding importance. Yet, if we are prepared to allow T_{max} to rise, we might be able to reduce $\bar{F}$ more than sufficiently to compensate. So let us turn our attention to minimising $\bar{F}$ subject to $T_{max} \leq \Delta$, i.e. subject to no job being finished more than Δ after its due date.

The key to our analysis is to realise that minimising $\bar{F}$ subject to $T_{max} \leq \Delta$ in one problem is equivalent to minimising $\bar{F}$ subject to $T_{max} = 0$ in another related problem in which all the due dates have been increased by Δ. A schedule has $T_{max} \leq \Delta$ in the first problem if and only if the same schedule has $T_{max} = 0$ in the second. Thus we may solve our $n/1//\bar{F}$ problem subject to $T_{max} \leq \Delta$ simply by adding Δ to all the due dates and applying Algorithm 4.4 to modified problem.

4.5 FINDING SCHEDULES EFFICIENT WITH RESPECT TO T_{max} AND $\bar{F}$

Let us turn now to the question of finding efficient schedules. Remember that a schedule is efficient with respect to T_{max} and $\bar{F}$ if we can find no other schedule at least as good in both criteria and strictly better in at least one (see condition (4.1)). We shall approach our task by considering the schedules constructed by Smith's algorithm. Are these efficient? In other words, if we minimise $\bar{F}$ subject to $T_{max} \leq \Delta$, do we find an efficient schedule? The answer, in short, is that we may do.

Consider what happens at Step 2 of the algorithm. As we have remarked, more than one job may satisfy condition (i) and (ii). At present, when this happens, we make an arbitrary choice of the job to process kth. Now a schedule that minimises $\bar{F}$ subject to $T_{max} \leq \Delta$ may, in fact, have $T_{max} < \Delta$. Moreover, the arbitrary choice at Step 2 may determine the final value of T_{max}. This suggests that, when two or more jobs satisfy condition (i) and (ii), we need to make a rather more careful choice. To argue what form this choice should take we need two observations: the first obvious, the second requiring a little thought.

(a) If we interchange two jobs not necessarily adjacent in the processing sequence, but both with the same processing time, then $\bar{F}$ is unchanged.

(b) Again suppose that there are two not necessarily adjacent jobs with the same processing time, but now suppose also that the job earlier in the sequence has the later due date. Then interchanging these two jobs cannot increase T_{max} and may, in fact, decrease it. To see this look back to the proofs of Theorem 3.4. There we were trying to minimise L_{max}, but we know that doing so also minimises T_{max} (Theorem 2.3). In that proof we interchanged two adjacent jobs. However, we only used the fact that they were adjacent at one point, namely in showing that the maximum lateness of the other $(n - 2)$ jobs was unaffected by the interchange. If the two jobs have the same processing time, then, whether or not they are adjacent, the maximum lateness of the other jobs will not be affected. Thus our observation here follows.

Let us now consider the choice that may face us at Step 2 of Smith's Algorithm. Consider any two jobs I and K satisfying (i) and (ii). By condition (ii) they must have the equal processing times. Suppose we choose job I and suppose that K has a later due date. Now K will be scheduled by some later cycle of the algorithm and so will be placed earlier in the processing sequence than I. By observation (a) we may interchange I and K in the processing sequence without affecting $\bar{F}$. Moreover, by observation (b) this interchange may decrease T_{max}. Thus it would have been better to select job K when making the choice of Step 2. Hence we modify Smith's Algorithm to ensure this choice. Step 2 is replaced by:

Step 2:   Find $J_{i(k)}$ in U such that (i) $d_{i(k)} \geq \tau$ and (ii) $p_{i(k)} \geq P_l$ for all J_l in U such that $d_l \geq \tau$. If there is a choice for $J_{i(k)}$, choose $J_{i(k)}$ to have the largest possible due date.

When this modification is made, the algorithm always finds an efficient schedule.

Theorem 4.5. Suppose that we use Smith's algorithm with the modified Step 2 to solve an $n/1//\bar{F}$ subject to $T_{max} \leq \Delta$ problem. Then either the algorithm fails to construct a schedule, i.e. it is impossible to have $T_{max} \leq \Delta$, or the resulting schedule is efficient with respect to T_{max} and $\bar{F}$. (N.B. it is understood that the algorithm is applied to the problem after all the due dates have been increased by Δ.)

Proof. We have already remarked that the algorithm will fail to construct a schedule if the constraint upon T_{max} is impossible. Suppose then it constructs a schedule S' with values T'_{max} and $\bar{F}'$. Let S be any other schedule with values T_{max} and $\bar{F}$. We consider three cases $\bar{F} < \bar{F}'$, $\bar{F} = \bar{F}'$, $\bar{F} > \bar{F}'$.

Suppose $\bar{F} < \bar{F}'$. Since $\bar{F}'$ is the minimum mean flow time subject to the restriction on the maximum tardiness, we must have $T'_{max} \leq \Delta < T_{max}$. Hence S does not dominate S'. (See (4.1).) Suppose $\bar{F} = \bar{F}'$. Note first that Theorem 4.3 gives an *if and only if* condition. So Smith's original algorithm can, in principle, find all the schedules that minimise mean flow time subject to the restriction on the maximum tardiness. Our modification to Step 2 ensures that of such schedules S' has the least possible maximum tardiness. Thus we must have $T'_{max} \leq T_{max}$ and again S does not dominate S'. Suppose $\bar{F} > \bar{F}'$. Then clearly S does not dominate S'.

In no case does S dominate S'. Hence S' must be efficient.

It should be noted that modification to Smith's algorithm may still leave us with a choice to make at Step 2. There may be two or more jobs, all satisfying conditions (i) and (ii) and all sharing the same due date. In such cases we can choose arbitrarily without any risk of upsetting the efficiency of the resulting schedule.

Our next task is to find all the efficient schedules for a problem. To do this we make one further assumption, namely that all the processing times and due dates are integral. Now for all possible schedules $T_{max} \leq \sum_{i=1}^{n} p_i$, since no job can be more tardy than the total processing time. Thus we may find all efficient schedules simply by minimising $\bar{F}$ subject to $T_{max} \leq \Delta$ for $\Delta = 0, 1, 2, \ldots, \sum_{i=1}^{n} p_i$. We need only consider integral values of Δ because we have assumed that the problem's data are integral. There are two qualifications that we should make immediately. First, we shall only find *all* efficient schedules if, whenever Step 2 involves an arbitrary choice between jobs with equal due dates, we repeat the solution until all possible sets of arbitrary choices have been made. Each repetition will lead to a distinct efficient schedule. Second, there is no need to solve the problems for all values of Δ. Suppose that in minimising $\bar{F}$ subject to $T_{max} \leq \Delta_1$ we obtain a schedule with $T_{max} = \Delta_0 < \Delta_1$. Then this same schedule will also minimize $\bar{F}$ subject to $T_{max} \leq \Delta$ for all Δ such that $\Delta_0 \leq \Delta \leq \Delta_1$. Thus we are led to the following algorithm which generates all the efficient schedules for our problem.

Algorithm 4.6 (Van Wassenhove and Gelders)

Step 1: Set $\Delta = \sum_{i=1}^{n} p_i$.

Step 2: Solve the $n/1//\bar{F}$ subject to $T_{max} \leq \Delta$ using the modified version of Smith's algorithm. If Step 2 of that algorithm involves an arbitrary choice, repeat the solution until all possible choices have been made. If there is no schedule with $T_{max} \leq \Delta$, go to *Step 5*.

Step 3: Let the schedule(s) found in *Step 2* have $T_{max} = \Delta_0$. Set $\Delta = \Delta_0 - 1$.

Step 4: If $\Delta \geq 0$, go to *Step 2*. Otherwise, continue to *Step 5*.

Step 5: Stop.

We consider an example taken directly from Van Wassenhove and Gelders (1980).

Example. Find all efficient schedules for the 4 job problem with data:

Job	J_1	J_2	J_3	J_4
Processing Time	2	4	3	1
Due Date	1	2	4	6

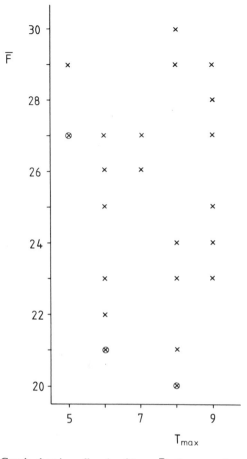

Fig. 4.3 Graph showing all pairs $(T_{max}, \bar{F})$ that may be attained by schedules for the example. $\otimes$ − $(T_{max}, \bar{F})$ pairs for the three efficient schedules. (Reproduced from Van Wassenhove and Gelders (1980) with permission.)

Applying Algorithm 4.6 we get:

Step 1: $\Delta = 10$.
Step 2: We find the efficient sequence (J_4, J_1, J_3, J_2) with $\bar{F} = 20$ and $T_{max} = 8$.
Step 3: $\Delta = 7$.
Step 4: $\Delta \geqslant 0$.
Step 2: We find the efficient sequence (J_4, J_1, J_2, J_3) with $\bar{F} = 21$ and $T_{max} = 6$.
Step 3: $\Delta = 5$.
Step 4: $\Delta \geqslant 0$.

Step 2: We find the efficient sequence (J_1, J_2, J_3, J_4) with $\bar{F} = 27$ and $T_{max} = 5$.

Step 3: $\Delta = 4$.

Step 4: $\Delta \geq 0$.

Step 2: No sequence with $T_{max} \leq 4$.

Step 5: Stop.

In Figure 4.3 we have plotted all the pairs $(T_{max}, \bar{F})$ that are obtainable by schedules in this example. The three schedules found above are clearly efficient, since no other schedules dominate them (see Figure 4.4)

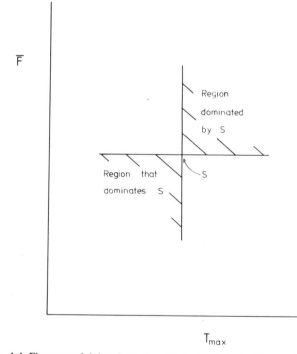

Fig. 4.4 Figure explaining how the efficient points in Figure 4.3 are recognised. The unshaded regions neither dominate S nor are dominated by S.

Suppose the total cost function in this example is linear with positive coefficients, say:

$$c(T_{max}, \bar{F}) = 4T_{max} + 7\bar{F}. \qquad (4.6)$$

Note that this $c(T_{max}, \bar{F})$ is increasing in both its arguments so, as we have argued in the introduction, the minimal total cost function must be efficient. The three efficient schedules have the following total costs.

(J_4, J_1, J_3, J_2): $T_{max} = 8$, $\bar{F} = 20 \Rightarrow$ total cost $= 4 \times 8 + 7 \times 20 = 172$.

(J_4, J_1, J_2, J_3): $T_{max} = 6$, $\bar{F} = 21 \Rightarrow$ total cost $= 4 \times 6 + 7 \times 21 = 171$

(J_1, J_2, J_3, J_4): $T_{max} = 5$, $\bar{F} = 27 \Rightarrow$ total cost $= 4 \times 5 + 7 \times 27 = 209$

Hence the minimal total cost schedule is (J_4, J_1, J_2, J_3).

There is an interesting and important geometric interpretation of this calculation, which will be familiar to those who have studied mathematical programming. (See Section 8.1 for references.) In Figure 4.5 we have plotted the family of lines given by $4T_{max} + 7\bar{F} = c$ for varying values of c. As c decreases these lines move across the figure as shown. Moreover, each line joins points $(T_{max}, \bar{F})$ that are equal in terms of total cost. The point $(6,21)$ corresponding to schedule (J_4, J_1, J_2, J_3) lies on the total cost line $4T_{max} + 7\bar{F} = 171$ and it is clear that no line of lower total cost passes through an efficient schedule. Thus we see immediately that this schedule achieves the minimum total cost.

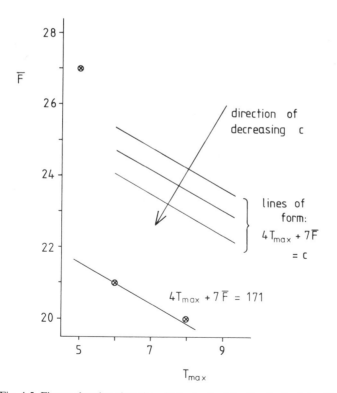

Fig. 4.5 Figure showing that the efficient schedule (J_4, J_1, J_2, J_3) with $T_{max} = 6$ and $\bar{F} = 21$ achieves the minimum total cost.

4.6 REFERENCES AND FURTHER READING

The section on required strings of jobs followed Conway, Maxwell and Miller (1967, pp. 70–71) fairly closely. Their book also considers single machine processing with more general forms of precedence constraints. However, for the most general treatment of $n/1//\bar{F}$ problems with precedence constraints, see Adolphson (1977). Lawler (1973) is the original source of the work in Section 4.3. Very recently Baker *et al.* (1980) have extended Lawler's approach to deal with non-zero ready times, but their solution does assume that pre-emption is allowed. The problem considered by Sidney (1977) has much in common with that studied by Lawler. However, the performance measure used by Sidney is specifically related to E_{max} and T_{max} and may fail the non-decreasing condition put on $\gamma_i(C_i)$ by Lawler. Glazebrook (1980) has considered single machine processing under precedence constraints when the costs are discounted over time.

The concept of efficiency has a far more general application in operational research than just in scheduling. It is also known by the terms **admissibility** and **Pareto optimality**. A brief, but excellent survey of the ideas and their importance is given in Keeney and Raiffa (1976, Chapter 3); a more theoretical and definitive study is given by Geoffrion (1968). Our treatment of efficiency with respect to T_{max} and $\bar{F}$ in single machine problems follows Van Wassenhove and Gelders (1980). Surprisingly, there does not seem to be a direct generalisation of these results to efficiency with respect to T_{max} and $\sum_{i=1}^{n} \alpha_i F_i$, i.e. a weighted mean flow time criterion. See Burns (1976) and Bansal (1980). Van Wassenhove and Baker (1980) introduce further ideas of efficiency into scheduling theory and solve another class of single machine problems. As apparent from the dates of these references, efficiency concepts have only just entered the scheduling literature. However, from their rapid adoption in other areas of operational research, we may expect substantial development of the topic.

4.7 PROBLEMS

1. Solve the following $12/1//\bar{F}$ problem given that jobs (3, 6, 9, 12) and jobs (1, 2, 4, 8, 10) must be processed in strings.

Job	1	2	3	4	5	6	7	8	9	10	11	12
Processing Time	2	1	5	6	3	4	1	2	5	6	1	6

2. Solve the $8/1//L_{max}$ problem:

Job	1	2	3	4	5	6	7	8
Processing Time	2	3	2	1	4	3	2	2
Due Date	5	4	13	6	12	10	15	19

with precedence constraints

$$2 \rightarrow 6 \rightarrow 3$$
$$1 \rightarrow 4 \rightarrow 7 \rightarrow 8$$

3. Solve the $4/1// \ \max_{i=1}^{4}\{\gamma_i(C_i)\}$ problem:

Job	1	2	3	4
Processing Time	2	1	2	1

with precedence constraints

$$2 \begin{array}{c} \nearrow 4 \\ \searrow 3 \end{array}$$

and where

$$\gamma_1(t) = t;$$
$$\gamma_2(t) = t - 2;$$
$$\gamma_3(t) = \begin{cases} (t-4)^2 & \text{if } t \geq 4, \\ 0 & \text{if } t < 4; \end{cases}$$
$$\gamma_4(t) = 2.\max\{0, t - 3\}.$$

4. Note that Lawler's method may be applied to problems without any precedence constraints. In particular, use Lawler's algorithm to show the EDD sequence solves both the $n/1//L_{max}$ and the $n/1//T_{max}$ problems.

5. Write down Lawler's Algorithm in an explicit step by step format.

6. Use Smith's Algorithm to solve the $7/1//\bar{F}$ subject to $T_{max} \leq 3$ problem with data:

Job	J_1	J_2	J_3	J_4	J_5	J_6	J_7
Processing Time	6	2	4	9	3	1	8
Due Date	33	13	6	22	31	38	14

7. Generate all schedules efficient with respect to T_{max} and $\bar{F}$ for the 7 job problem in question 6. Hence find the minimum total cost schedule for $c(T_{max}, \bar{F}) = 2T_{max} + 7\bar{F}.$

Chapter **5**

Constructive Algorithms for Flow-Shops and Job-Shops

5.1 INTRODUCTION

It is now time to turn our attention to problems with more than one machine, i.e. scheduling in the flow-shop or general job-shop. Our immediate concern will be to study problems for which **constructive algorithms** exist. By a constructive algorithm we mean one which builds up an optimal solution from the data of the problem by following a simple set of rules which exactly determine the processing order. In the single machine case we encountered a variety of problems which could be solved constructively. We shall not be so fortunate here; only a very few with two or more machines are amenable to such analysis. Indeed, the $n/2/G/F_{max}$ family of problems is the only one for which there exists a constructive algorithm applicable to all cases. For all other families the few constructive algorithms that exist apply only to special cases, usually cases in which the processing times are restricted in some way. In this chapter we consider the most important of these algorithms, namely those due to Johnson (1954).

We make only one assumption over and above those made in Chapter 1. We assume that all jobs are available for processing immediately. Thus:

Assumption for Chapter 5
(i) $r_i = 0$ for all J_i, $i = 1, 2, \ldots, n$.

5.2 SOME RESULTS ON SCHEDULES IN A FLOW-SHOP

In this section we consider two very general results about the form of schedules in a flow-shop. In a flow-shop the technological constraints demand that the jobs pass between the machines in the same order; i.e. if J_1 must be processed on machine M_k before machine M_l, then the same is true for all jobs. We shall see that these general results imply that for some 2 and 3 machine flow-shops we need only consider permutation schedules. A

permutation schedule is one on which each machine processes the jobs in the same order; i.e. if on machine M_1 job J_i is processed before J_k, then the same is true for all machines.

In flow-shops there is a natural ordering of the machines, namely that given by the technological constraints as the processing order for each and every job. Thus we shall assume that each job must be processed by M_1 before M_2 before M_3 etc. The technological constraints, therefore, have the form:

Table 5.1—The technological constraints in a flow-shop

Job	Processing order				
J_1	M_1	M_2	M_3	...	M_m
J_2	M_1	M_2	M_3	...	M_m
.	.	.	.		.
.	.	.	.		.
.	.	.	.		.
J_n	M_1	M_2	M_3	...	M_m

Theorem 5.1. For the $n/m/F/B$ problem with B a regular measure of performance, it is sufficient to consider schedules in which the same processing sequence is given on the first two machines.

Proof. If a schedule S_1 does not have the same order on both machines M_1 and M_2 there is a job I which directly precedes job K on M_1, but follows K, perhaps with intervening jobs, on M_2. (See Fig. 5.1(a).) On M_1 we may reverse the order of I and K without increasing any starting time on M_2. Thus this interchange cannot increase the completion time of any job and, hence, neither can it increase a regular measure.

This process of exchanging jobs on M_1 may be repeated until a schedule S_1' is obtained with the same order on M_1 as on M_2. Clearly S_1' is no worse than S_1 under a regular measure.

Theorem 5.2. For the $n/m/F/F_{\max}$ problem, there is no need to consider schedules with different processing orders on machines M_{m-1} and M_m.

Proof. If a schedule S_2 does not have the same order on both machines M_{m-1} and M_m, there is a job I which directly precedes job K on M_{m-1}, but follows K, perhaps with some intervening jobs, on M_m. (See Fig. 5.1(b).) Suppose we reverse the processing order of I and K on M_m. Clearly this may change the flow times of individual jobs: some may increase; some

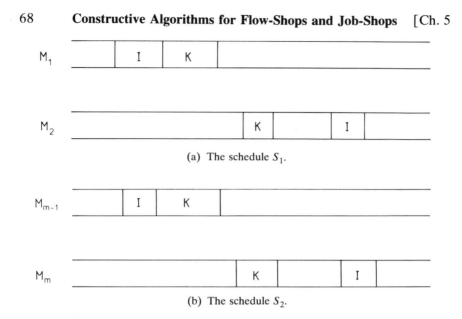

(a) The schedule S_1.

(b) The schedule S_2.

Fig. 5.1 The schedules used in proving Theorems 5.1 and 5.2.

may decrease. However, in total the processing on M_m can only be expedited, because I completes on M_{m-1} before K does. Thus F_{max} cannot be increased by the interchange.

This process of interchanging jobs on M_m may be repeated until a schedule S_2' is obtained with the same order on M_{m-1} as on M_m. The above shows that S_2' can be no worse than S_2 in terms of F_{max}.

The proofs of Theorems 5.1 and 5.2 are very similar, but by no means the same. The similarity is evident from Fig. 5.1(a) and (b); these differ only in the labelling of the machines. The important distinction between the proofs lies in the machine chosen for the interchange of processing order. In the proof of Theorem 5.1 we reversed the order of I and K on M_1, the first of the two machines considered. We saw that doing so could improve and certainly not worsen each and every flow time. Ideally we should have liked to copy this interchange in the proof of Theorem 5.2, i.e. to have interchanged jobs on M_{m-1}. For then that theorem would have been applicable to problems of scheduling against any regular measure and not just F_{max}. However, a little thought shows that the effect of interchanging jobs on M_{m-1} is entirely unpredictable; what happens depends upon the times jobs complete on M_{m-2}. Thus we must make the interchange on M_m and so can only deduce a result in terms of F_{max}.

Because of Theorem 5.1 we need only consider permutation schedules for the $n/2/F/B$ problem, when B is regular. Adding to this the result of

Theorem 5.2, again we need only consider permutation schedules for the $n/3/F/F_{max}$ problem. Examples given in Problems 5.8.1 show that these results cannot be strengthened to more machines.

5.3 JOHNSON'S ALGORITHM FOR THE $n/2/F/F_{max}$ PROBLEM

Suppose we are faced with an $n/2/F/F_{max}$ problem; i.e. we have to process n jobs through two machines, each job in the order M_1, M_2, so that the maximum flow time is minimized. Note that, because all jobs have zero ready times, $F_{max} = C_{max}$. Now it seems sensible to start the processing with the job with the shortest processing time on M_1. For then the processing on M_2 may start as soon as possible. (See Fig. 5.2.) Similarly it seems just as sensible to finish the processing with the job that has the shortest processing time on M_2. For while this job is processing M_1 must be idle. Also we have just shown that we only need to consider permutation schedules for this problem. Putting these ideas together, it seems reasonable to suggest that the optimal schedule is a permutation of $\{J_1, J_2, \ldots, J_n\}$ such that the earlier jobs in the processing sequence have short M_1 processing times, whereas the later jobs have short M_2 processing times. The schedule constructed by Johnson's Algorithm has precisely this property.

We shall give Johnson's Algorithm next and an example of its use. Then we shall prove that his method does find an optimal schedule. It will simplify our notation to let:

$$a_i = p_{i1}, \text{ the processing time of the } J_i \text{ on } M_1;$$

and

$$b_i = p_{i2}, \text{ the processing time of the } J_i \text{ on } M_2.$$

The algorithm builds up the processing sequence by working in from the ends towards the middle. To do this we shall need two counters, namely k

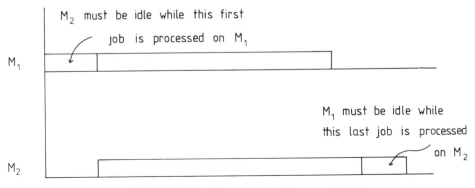

Fig. 5.2 The idle time that must occur for any schedule.

and l. $k = 1$ initially and increases 2, 3, 4 . . . as the 1st, 2nd, 3rd, 4th . . . positions in the processing sequence are filled. Similarly, $l = n$ initially and decreases $(n - 1)$, $(n - 2)$, . . . as the nth, $(n - 1)$th, $(n - 2)$th, . . . positions in the processing sequence are filled.

Algorithm 5.3 (Johnson)

Step 1 Set $k = 1, l = n$.
Step 2 Set current list of unscheduled jobs = $\{J_1, J_2, \ldots, J_n\}$.
Step 3 Find the smallest of all the a_i and b_i times for the jobs currently unscheduled.
Step 4 If the smallest time is for J_i on first machine, i.e. a_i is smallest, then
  (i) Schedule J_i in kth position of processing sequence.
  (ii) Delete J_i from current list of unscheduled jobs.
  (iii) Increment k to $k + 1$.
  (iv) Go to *Step 6*.
Step 5 If the smallest time is for J_i on second machine, i.e. b_i is smallest, then:
  (i) Schedule J_i in the lth position of processing sequence.
  (ii) Delete J_i from current list of unscheduled jobs.
  (iii) Reduce l to $(l - 1)$.
  (iv) Go to *Step 6*.
Step 6 If there are any jobs still unscheduled, go to *Step 3*. Otherwise, stop.

N.B. If the smallest time occurs for more than one job in Step 3, then pick J_i arbitrarily.

Example. Schedule the $7/2/F/F_{max}$ problem with data:

Job	Processing Time on Machine M_1	M_2
1	6	3
2	2	9
3	4	3
4	1	8
5	7	1
6	4	5
7	7	6

Applying the algorithm the schedule builds up as follows.

```
Job 4 scheduled:  4  _ _ _ _ _ _
Job 5 scheduled:  4  _ _ _ _ _ 5
Job 2 scheduled:  4 2 _ _ _ _ 5
Job 3 scheduled:  4 2 _ _ _ 3 5
Job 1 scheduled:  4 2 _ _ 1 3 5
Job 6 scheduled:  4 2 6  1 3 5
Job 7 scheduled:  4 2 6 7 1 3 5
```

Thus we should sequence the jobs in the order (4, 2, 6, 7, 1, 3, 5).

N.B. In the above there are two arbitrary choices. We could have put Job 5 into the last position of the sequence before scheduling Job 4. Here the resulting sequence would have been the same. Also we could have scheduled Job 1 in the sixth position instead of Job 3. This would have lead to a different, but equivalent, processing sequence, viz. (4, 2, 6, 7, 3, 1, 5).

We now prove that Johnson's Algorithm does produce an optimal schedule. We begin by showing that the first cycle of the algorithm chooses and positions a job optimally.

Theorem 5.4. For the $n/2/F/F_{\max}$ problem with $p_{i1} = a_i$ and $p_{i2} = b_i, i = 1, 2, \ldots, n$:

    (i) If $a_k = \min\{a_1, a_2, \ldots, a_n, b_1, b_2, \ldots, b_n\}$, there is an optimal schedule with J_k first in the processing sequence;

    (ii) If $b_k = \min\{a_1, a_2, \ldots, a_n, b_1, b_2, \ldots, b_n\}$, there is an optimal schedule with J_k last in the processing sequence.

Proof. We prove (i) first. Let J_k be such that $a_k = \min\{a_1, a_2, \ldots, a_n, b_1, b_2, \ldots, b_n\}$. Let S be a schedule with J_k not first in the processing sequence. Let J_l be the job which is processed immediately before J_k. Then S has a Gantt diagram as given in Fig. 5.3. Let Q be the time that M_1 is ready to start processing J_l under S. Let R be the time that M_2 finishes processing the job before J_l under S and is thus ready to process J_l.

We consider the schedule S' obtained simply by interchanging the jobs J_k and J_l in S. Our aim is to show that $C_k \geqslant C'_l$, where C_k is the completion time of J_k under S and C'_l is the completion time of J_l under S'. If this is true $F_{\max} = C_{\max} \geqslant C'_{\max} = F'_{\max}$ and we deduce that (1) holds by repeatedly interchanging J_k with the jobs before it until it is first to be processed.

Under S, J_l starts on M_2 at $\max\{R, Q + a_l\}$. We know, by assumption, that $a_k \leqslant b_l$. Thus J_l completes on M_2 after J_k completes on M_1 and so we have:

$$C_k = \max\{R, Q + a_l\} + b_l + b_k.$$

Under S', J_k starts on M_2 at $\max\{R, Q + a_k\}$ and J_l starts on M_2 either as soon as J_k finishes or as soon as M_1 finishes processing J_l, whichever is the

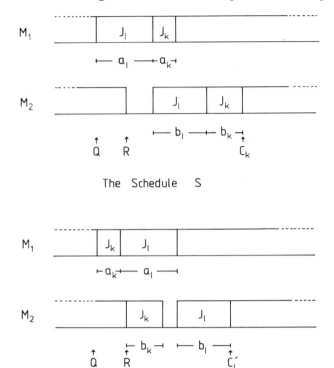

Fig. 5.3 The schedules S and S'.

sooner. So

$$C'_l = \max\{\max\{R, Q + a_k\} + b_k, Q + a_k + a_l\} + b_l.$$

Hence $$C'_l = \max\{\max\{R, Q + a_k\} + b_k + b_l, Q + a_k + a_l + b_l\}. \qquad (5.1)$$

Now $$C_k = \max\{R, Q + a_l\} + b_l + b_k$$
$$\geq \max\{R, Q + a_k\} + b_l + b_k, \text{ because } a_l \geq a_k \text{ (by assumption)}.$$

and $$C_k = \max\{R, Q + a_l\} + b_l + b_k$$
$$\geq Q + a_l + b_l + b_k$$
$$\geq Q + a_l + b_l + a_k, \qquad\qquad \text{because } b_k \geq a_k \text{ (by assumption)}.$$

So C_k is no less than either of the terms in the outer max{ } of (5.1). Thus $C_k \geq C'_l$ as required.

The proof of (ii) is left for Problem 5.8.2.

It is now a relatively simple matter to deduce that Johnson's Algorithm also positions the remaining $(n - 1)$ jobs correctly.

Theorem 5.5. Algorithm 5.3 finds an optimal schedule for the $n/2/F/F_{max}$ problem.

Proof. By Theorem 5.4 we know that the first cycle of the algorithm positions a job optimally. We shall show that, if the first v cycles have positioned v jobs optimally, the $(v + 1)$st cycle will also position a job optimally. Our result then follows by induction.

Suppose v cycles have positioned v jobs optimally. Because of this optimality the $(v + 1)$st cycle need not reconsider any of the choices made in earlier cycles; it may simply choose one of the remaining $(n - v)$ jobs and position it in one of the remaining $(n - v)$ places in the processing sequence. What the cycle actually does is to choose and position a job according to the conditions of Theorem 5.4 applied to the remaining $(n - v)$ unscheduled jobs alone. Look again at the proof of that theorem. The argument shows that $C_k \geqslant C_l'$ for the interchanged jobs. Moreover, only quantities directly relating to J_k and J_l enter the argument; the rest of the processing sequence is irrelevant. It follows that, in placing the $(v + 1)$st job to minimise F_{max} for the remaining $(n - v)$ jobs considered alone, the algorithm is minimising F_{max} for all n jobs. Thus the $(v + 1)$st cycle positions a job optimally.

We know that the first cycle positions a job optimally. Setting $v = 1$, the above shows that the second cycle also positions a job optimally. Next setting $v = 2$ we deduce that the third cycle does so too. Continuing in this way, we deduce that all cycles position jobs optimally. Thus the algorithm constructs an optimal schedule.

5.4 JOHNSON'S ALGORITHM FOR THE $n/2/G/F_{max}$ PROBLEM

In Section 1.4 we made the assumption that each job must be processed through all the machines. For this section we drop that assumption.

Suppose that the set of n jobs $\{J_1, J_2, \ldots, J_n\}$ may be partitioned into four types of job as follows.

Type A: Those to be processed on machine M_1 only.
Type B: Those to be processed on machine M_2 only.
Type C: Those to be processed on both machines in the order M_1 then M_2.
Type D: Those to be processed on both machines in the order M_2 then M_1.

Then a little thought shows that the construction of an optimal schedule is straightforward.

(1) Schedule the jobs of type A in any order to give the sequence S_A.
(2) Schedule the jobs of type B in any order to give the sequence S_B.
(3) Schedule the jobs of type C according to Johnson's Algorithm for $n/2/F/F_{max}$ problems to give the sequence S_C.

(4) Schedule the jobs of type D according to Johnson's Algorithm for $n/2/F/F_{max}$ problems to give the sequence S_D. (*Beware:* Here M_2 is the first machine and M_1 the second in the technological constraints.)

An optimal schedule is then:

Machine	Processing Order
M_1	(S_C, S_A, S_D)
M_2	(S_D, S_B, S_C)

To see that this schedule is optimal remember that time is wasted and hence F_{max} is increased, if either M_2 is kept idle waiting for jobs of type C to complete on M_1 or M_1 is kept idle waiting for jobs of type D to complete on M_2. This schedule clearly minimises such idle time.

Example. Consider the $9/2/G/F_{max}$ problem with times and processing order given by:

Job	Processing order and times			
	First Machine		Second Machine	
1	M_1	8	M_2	2
2	M_1	7	M_2	5
3	M_1	9	M_2	8
4	M_1	4	M_2	7
5	M_2	6	M_1	4
6	M_2	5	M_1	3
7	M_1	9	–	
8	M_2	1	–	
9	M_2	5	–	

To find an optimal schedule:

Type A jobs – only job 7 is to be processed on M_1 alone.
Type B jobs – jobs 8 and 9 require M_2 alone. Select arbitrary order (8, 9).
Type C jobs – jobs 1, 2, 3 and 4 require M_1 first and then M_2. Johnson's Algorithm for this $4/2/F/F_{max}$ problem gives the sequence (4, 3, 2, 1).
Type D jobs   jobs 5, 6 require M_2 first and then M_1. Johnson's Algorithm for this $2/2/F/F_{max}$ problem gives the sequence (5, 6) (Remember M_1 is now the *second* machine).

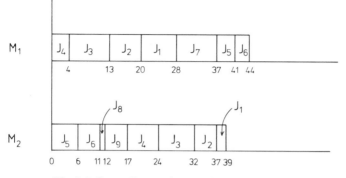

Fig. 5.4 Gantt diagram for the $9/2/G/F_{\max}$ example.

Thus an optimal sequence for the overall problem is

Processing Sequence

Machine M_1	(4, 3, 2, 1, 7, 5, 6)
Machine M_2	(5, 6, 8, 9, 4, 3, 2, 1)

and the resulting Gantt diagram is given in Fig. 5.4. From this we see that $F_{\max} = 44$ for an optimal schedule.

5.5 A SPECIAL CASE OF THE $n/3/F/F_{\max}$ PROBLEM

Johnson's Algorithm for the $n/2/F/F_{\max}$ problem may be extended to a special case of the $n/3/F/F_{\max}$ problem. We need the condition that:

$$\text{either} \qquad \min_{i=1}^{n}\{p_{i1}\} \geq \max_{i=1}^{n}\{p_{i2}\}$$

$$\text{or} \qquad \min_{i=1}^{n}\{p_{i3}\} \geq \max_{i=1}^{n}\{p_{i2}\}$$

(5.2)

i.e. the maximum processing time on the second machine is no greater than the minimum time on either the first or the third. If (5.2) holds, an optimal schedule for the problem may be found by letting

$$a_i = p_{i1} + p_{i2},$$

$$b_i = p_{i2} + p_{i3},$$

and scheduling the jobs as if they are to be processed on two machines only, but with the processing time of each job being a_i and b_i on the first and second machines respectively.

Remember we have shown that we need only consider permutation schedules for the $n/3/F/F_{max}$ problem and note that Johnson's Algorithm applied to the $n/2/F/F_{max}$ problem with the constructed a_i and b_i times does produce such a schedule. I ask you to prove in Problem 5.8.8 that it is, indeed, an optimal schedule. In Section 7.3 we encounter an example, which shows that condition (5.2) is necessary for this method to find an optimal schedule.

Example. Consider the $6/3/F/F_{max}$ problem with data:

Job	Actual Processing Times			Constructed Processing Times	
	M_1	M_2	M_2	1st Machine	2nd Machine
1	4	1	3	5	4
2	6	2	9	8	11
3	3	1	2	4	3
4	5	3	7	8	10
5	8	2	6	10	8
6	4	1	1	5	2

First we check that (5.2) holds for this problem. Here we have

$$\min_{i=1}^{6}\{p_{1i}\} = 3; \quad \max_{i=1}^{6}\{p_{2i}\} = 3; \quad \text{and} \quad \min_{i=1}^{6}\{p_{i3}\} = 1.$$

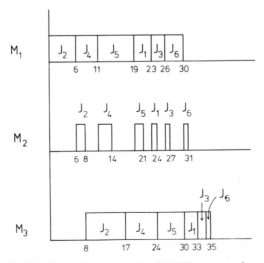

Fig. 5.5 Gantt diagram for the $6/3/F/F_{max}$ example.

Thus we have $\min_{i=1}^{6}\{p_{1i}\} = 3 \geq 3 = \max_{i=1}^{6}\{p_{2i}\}$ and (5.2) holds. (N.B. (5.2) is an either-or condition; we do not need both inequalities to hold.)

The constructed a_i and b_i times are given in the table. Applying Algorithm 5.3 gives the sequence (2, 4, 5, 1, 3, 6), which has the Gantt diagram given in Fig. 5.5.

5.6 AKERS' GRAPHICAL SOLUTION TO THE $2/m/F/F_{max}$ PROBLEM

We have now examined all the constructive methods for producing optimal schedules that we shall in this book. The algorithms in the remaining chapters are enumerative ones. They generate schedules one by one, searching for an optimal solution. Often, as we shall see, they use 'clever' elimination procedures to see if the non-optimality of one schedule implies the non-optimality of many others not yet generated. Thus the methods may not search all the set of feasible solutions. Nonetheless, they are methods that proceed by exhaustive enumeration and, hence, require much computation. As a first step in this move to the study of enumeration methods, we consider a graphical method due to Akers (1956).

We consider a 2 job, 6 machine flow-shop problem $2/6/F/F_{max}$. The case for general m machine flow-shops follows accordingly. In Fig. 5.6 the x-axis is marked off in intervals corresponding to the processing times of

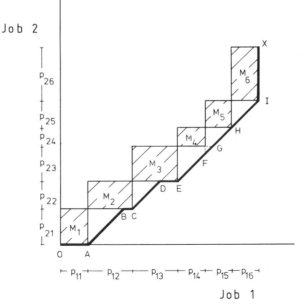

Fig. 5.6 Graph of a schedule in a $2/6/F/F_{max}$ problem.

job 1. Similarly the y-axis is marked off in intervals corresponding to those of job 2. Note that the machines have a natural ordering since this is a flow-shop problem. The rectangles whose sides correspond to the processing times of jobs 1 and 2 on the same machine are shaded off. These will correspond to 'forbidden zones', meaning that any one machine cannot process both jobs simultaneously.

A schedule may be represented on this diagram by plotting a line that

(a) runs from O to X, i.e. from $(0, 0)$ to $(\Sigma_{j=1}^{6} p_{1j}, \Sigma_{j=1}^{6} p_{2j})$;

(b) is composed of segments running

   (i) horizontally, representing work on job 1 only,

   (ii) vertically, representing work on job 2 only,

   (iii) at 45°, representing work on both jobs;

(c) does not enter any forbidden zone, i.e. does not demand that any machine process both jobs simultaneously.

The easiest way to understand this is to consider a particular example: the schedule S represented by the line OABCDEFGHIX in Figure 5.6. Its processing is described in Table 5.2.

Table 5.2—Description of the schedule S shown in Figure 5.6

SEGMENT	Job 1	Job 2
O A	being processed on M_1	waiting
A B	being processed on M_2	being processed on M_1
B C	being processed on M_2	waiting
C D	being processed on M_3	being processed on M_2
D E	being processed on M_3	waiting
E F	being processed on M_4	being processed on M_3
F G	being processed on M_5	being processed on M_3
G H	being processed on M_5	being processed on M_4
H I	being processed on M_6	being processed on M_5
I X	completed	being processed on M_6

Notice that the schedule S is a permutation schedule; all machines process the jobs in the order $(1,2)$. It will be worth your while to stop and interpret the schedule S' given by the line OABCDEFGHIJX in Fig. 5.7. Note that S' is not a permutation schedule.

The maximum flow time of a schedule is given by either

F_{max} = total processing time on job 1
   + total time during which job 2 is processed while job 1 is waiting

$$= \sum_{j=1}^{6} p_{1j} + \text{sum of lengths of vertical segments of schedule line,}$$

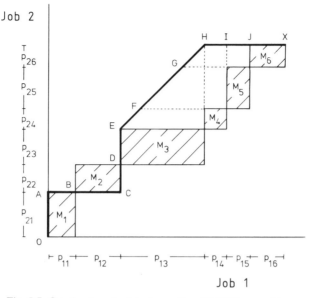

Fig. 5.7 Graph of a schedule in another $2/6/F/F_{max}$ problem.

or

F_{max} = total processing time on job 2
+ total time during which job 1 is processed while job 2 is waiting

$$= \sum_{j=1}^{6} p_{2j} + \text{sum of lengths of horizontal segments of schedule line.}$$

The terms $\sum_{j=1}^{6} p_{1j}$ and $\sum_{j=1}^{6} p_{2j}$ are independent of the schedule. Thus to find an optimal schedule we must construct the schedule line that has the least length of vertical segments or, equivalently, the least length of horizontal segments. How do we construct these?

Essentially each schedule line represents a choice for each shaded forbidden region. It may either

(a) pass above the region indicating that job 2 is processed on that machine before job 1,

or (b) pass below the region indicating that job 1 is processed on that machine before job 2.

These choices are represented in Fig. 5.8, which shows the configurations which may result. In each of the four cases a choice must be made at the point C. Should job 2 be processed on M_j before job 1, in which case the upper branch must be taken? Or should job 1 be processed first and thus the lower branch followed? Note that in cases (b) and (d) the choice is obvious. In (b) job 1 can complete on M_j before job 2 requires it. Thus,

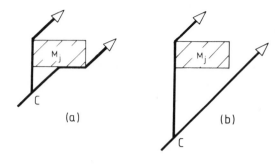

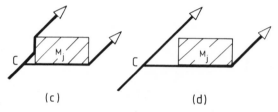

Fig. 5.8 Possible branches at each forbidden region.

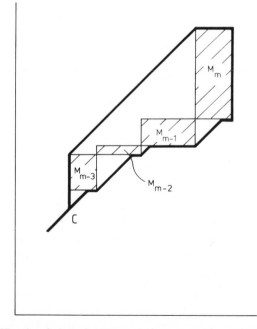

Fig. 5.9 Why the obvious short term choice may not be optimal in the
long term.

even if later in the schedule there is an advantage in holding up job 1 to let job 2 overtake it, there is no need to consider that yet and the lower branch should be taken. (See Problem 5.8.10.) Similarly, in case (d) the upper branch should be taken.

This construction may give 2^m possible schedules or, rather, 2^{m-2} if we also make use of Theorems 5.1 and 5.2. Thus for a problem with 12 machines we may need to consider $2^{10} = 1024$ schedule lines. However, in practice, cases (b) and (d) happen a lot, and also the human eye is very good at spotting the schedule with the least total length of horizontal segments. Thus it is normally fairly easy to find an optimal schedule simply by looking at the graph.

Finally, we note an example which indicates the complexity of the scheduling task in general. Consider Fig. 5.9, which shows the last four machines in a $2/m/F/F_{max}$ problem. At point C it seems attractive in the short term to pass under the $M_{(m-3)}$ forbidden zone. However, doing so 'traps' the schedule in a region with little possibility of using 45° lines. Passing over the $M_{(m-3)}$ forbidden zone is clearly optimal. In other words, when deciding the processing order of a pair of jobs on a particular machine, it is necessary to consider the ramifications of this choice for the scheduling of all other machines. Scheduling is generally a very difficult task because few, if any, of one's choices are independent of all the others.

5.7 REFERENCES AND FURTHER READING

Johnson (1954) is the original source of the material presented in Sections 5.1–5.4; his paper has been reprinted in Muth and Thompson (1963). There are many alternative proofs of his results: Conway, Maxwell, and Miller (1967) and Baker (1974) generally follow Johnson's approach; Rinnooy Kan (1976) gives entirely different proofs, upon which ours are based; and White (1969) uses dynamic programming ideas to derive the same results.

Szwarc (1977) has examined the $n/3/F/F_{max}$ problem in detail, extending and enlarging upon Johnson's results. General conditions for optimality have been obtained for the general m machine flow-shop by Yueh Minh-I (1976), but his results lead to no useful algorithms. Panwalker and his co-workers have produced constructive algorithms for several very special cases of the m-machine flow-shop problem in which the processing times are restricted to obey many conditions on their relative magnitude. (Panwalker and Khan (1976); Smith *et al* (1976); Panwalker and Woollam (1979, 1980)).

The presentation of Akers graphical method given in the previous section closely follows that of Conway, Maxwell and Miller (1967, pp. 98–100). Because of its graphical nature, it is a method confined to two dimensions and hence does not generalise obviously.

5.8 PROBLEMS

1.(a) Consider the $2/3/F/\bar{F}$ problem with processing times given below.

Machine

Job	M_1	M_2	M_3
J_1	5	1	1
J_2	1	5	5

Show that the two permutation schedules (J_1, J_2) and (J_2, J_1) both have mean flow times of 11.5, and that the schedule below has a mean flow time of 10.5.

Machine	Processing Order	
M_1	J_2	J_1
M_2	J_2	J_1
M_3	J_1	J_2

(b) Consider the $2/4/F/F_{max}$ problem with processing times:

Machine

Jobs	M_1	M_2	M_3	M_4
J_1	5	1	1	5
J_2	1	5	5	1

Show that both permutation schedules have $F_{max} = 17$, whereas the schedule below has $F_{max} = 14$.

Machine	Processing Order	
M_1	J_2	J_1
M_2	J_2	J_1
M_3	J_1	J_2
M_4	J_1	J_2

2. Prove Theorem 5.4(ii).

3. Schedule the $8/2/F/F_{max}$ problem with processing times:

Job	Machine	
	M_1	M_2
1	2	10
2	3	7
3	8	2
4	7	5
5	9	3
6	4	4
7	1	9
8	6	8

Find the maximum flow time for the optimal schedule.

4. By considering the following processing times show that Johnson's Algorithm does not solve the $n/2/F/\bar{F}$ problem:

Job	Machine	
	M_1	M_2
1	1	7
2	1	$\frac{3}{4}$
3	1	$\frac{3}{4}$
4	1	$\frac{3}{4}$
5	1	$\frac{3}{4}$
6	1	$\frac{3}{4}$

5. Generalise Problem 5.8.4 as below in order to show that Johnson's algorithm may be very poor for the $n/2/F/\bar{F}$ problem. Consider n jobs with processing times $a_i = \varepsilon$ for $i = 1, 2, 3, \ldots n$; $b_1 = \beta$; and $b_i = \frac{3}{4}\varepsilon$ for $i = 2, 3, \ldots, n$. Assume $n\varepsilon < \beta$. Show firstly that Johnson's Algorithm may give the processing sequence $(J_1, J_2, J_3, \ldots, J_n)$. For this sequence show that the mean flow time is

$$\bar{F}_{Johnson} = \beta + \varepsilon + (n - 1)3\varepsilon/8.$$

Next show that an optimal sequence is $(J_2, J_3, \ldots, J_n, J_1)$.
[*Hint:* Show that $\bar{F}$ only depends on the position of J_1 in the sequence and find this dependence]. For this optimal sequence show

$$\bar{F}_{opt} = \beta/n + (n + 1)\varepsilon/2 + \frac{3(n - 1)\varepsilon}{4n}$$

Hence show, that as $\varepsilon \to 0$, $\overline{F}_{\text{Johnson}}/\overline{F}_{\text{opt}} \to n$.

This example is due to Ganzalez and Sahni (1978).

6. Solve the $14/2/G/F_{\text{max}}$ problem below and find the optimal maximum flow time.

	Processing Order and Time			
Job	**First Machine**		**Second Machine**	
1	M_1	8	M_2	7
2	M_2	12	M_1	14
3	M_1	13	M_2	9
4	M_1	6	–	–
5	M_1	7	M_2	13
6	M_2	9	M_1	8
7	M_2	12	–	–
8	M_1	4	M_2	10
9	M_1	10	M_2	6
10	M_2	10	M_1	11
11	M_2	6	–	–
12	M_2	8	–	–
13	M_1	5	M_2	13
14	M_1	7	–	–

7. Solve the $6/3/F/F_{\text{max}}$ problem and draw a Gantt diagram to find the optimal F_{max}

	Processing Times		
Job	**M_1**	**M_2**	**M_3**
1	10	9	10
2	2	8	10
3	3	7	10
4	6	6	10
5	1	5	10
6	4	4	10

8. Prove that Johnson's solution of the $n/3/F/F_{\text{max}}$ problem, when condition (5.2) holds, does give an optimal schedule.

Hint: Let S be any schedule and assume the jobs are renumbered in this order, i.e. $S = (1, 2, 3, \ldots, n)$. Let $p_{i1} = \alpha_i$, $p_{i2} = \beta_i$, and $p_{i3} = \gamma_i$ for $i = 1, 2, \ldots, n$. Assume $\min_{i=1}^{n}\{\alpha_i\} \geq \max_{i=1}^{n}\{\beta_i\}$. Also let s_{ij} be the starting time of J_i on machine j, $j = 1, 2, 3$. Show that

$$s_{i1} = s_{(i-1)1} + \alpha_{i-1} \qquad i = 2, 3, \ldots, n$$
$$s_{i2} = s_{i1} + \alpha_i \qquad i = 1, 2, 3, \ldots, n$$

$$s_{i3} = \max\{s_{(i-1)3} + \gamma_{i-1}, s_{i1} + \alpha_i + \beta_i\} \; i = 2, 3, \ldots, n$$

Now let $a_i = \alpha_i + \beta_i$ and $b_i = \beta_i + \gamma_i$ and consider the schedule S for the $n/2/F/F_{max}$ problem with these constructed processing times. Let t_{ij} be the starting time of J_i on machine j for this 2 machine problem under S.

Show that $\qquad t_{i1} = s_{i1} + \displaystyle\sum_{l=1}^{i-1} \beta_l$

$$i = 1, 2, 3 \ldots, n.$$

and that $\qquad t_{i2} = s_{i3} + \displaystyle\sum_{l=1}^{i-1} \beta_l$

Hence deduce that maximum flow time under S for original problem = maximum flow time under S for constructed problem $- \sum_{l=1}^{n}\beta_l$. Since the term $\sum_{l=1}^{n}\beta_l$ is independent of S, minimising the maximum flow time for the constructed problem also minimises it for the original. The proof for the case where $\min_{i=1}^{n}\{\gamma_i\} \geq \max_{i=1}^{n}\{\beta_i\}$ is similar.

9. Explain carefully the interpretation of the schedule S' in Fig. 5.7. Why would one not need to consider this schedule in looking for an optimal schedule?

10. Show, by considering Fig. 5.10 that in case (c) of Fig. 5.8 the lower path may always be chosen at the point C.

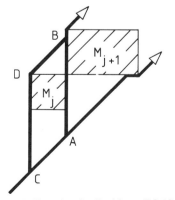

Fig. 5.10 Situation for Problems 5.8.10.

11. Solve the $2/6/F/F_{max}$ problem with processing times:

	Processing Times	
Machine	**Job 1**	**Job 2**
M_1	4	3
M_2	6	7
M_3	2	4
M_4	5	2
M_5	3	5
M_6	8	6

Chapter **6**

Dynamic Programming Approaches

6.1 INTRODUCTION

As remarked in the last chapter, we have now completed our discussion of constructive methods for solving scheduling problems and we pass on to consider **enumeration** methods. These simply list, or enumerate, all possible schedules and then eliminate the non-optimal schedules from the list, leaving those which are optimal. In this chapter we discuss one such general approach called **dynamic programming.**

Dynamic programming originates from Bellman (1957) and is applicable to very many optimising problems, not just those arising in scheduling. White (1969) provides an introductory treatment and an extensive review of its applications; also most general operational research textbooks discuss it briefly. Roughly, the method applies to any problem which can be broken down into a sequence of nested problems, the solution of one being derived in a straight-forward fashion from that of the preceding problem. However, we shall not discuss the dynamic programming approach in any great generality, but rather look at a specific application in the field of scheduling, and allow, I hope, our intuition to abstract the underlying ideas.

6.2 THE APPROACH OF HELD AND KARP

In 1962 Held and Karp followed the lead of Bellman and applied dynamic programming ideas to sequencing problems. Their method applies to single machine problems where the performance measure takes the form: $\sum_{i=1}^{n} \gamma_i(C_i)$. Here the $\gamma_i(C_i)$ are assumed to be non-decreasing functions of the completion times. It follows immediately that this performance measure is regular. As particular instances we note that $\bar{C}$, $\bar{F}$ and $\bar{T}$ take this form. (Set $\gamma_i(C_i)$ equal to C_i/n, $(C_i - r_i)/n$, and $\max\{C_i - d_i, 0\}/n$ respectively.) Thus we may use Held and Karp's method to solve $n/1//C, n/1//F$, and $n/1//\bar{T}$ problems. Of course, it would be foolish to use it on the first two problems; we know that an SPT schedule solves those. But we have been unable to solve $n/1//\overline{T}$ problems until now. The approach may also be extended to solve problems with performance measures of the form $\{\max_{i=1}^{n} \gamma_i(C_i)\}$. (See Problem 6.6.6). However, again dynamic programming is not the best method of solution; Lawler's Algorithm is.

Held and Karp's approach is based upon a simple observation about the structure of an optimal schedule. Roughly, this says that in an optimal schedule the first K jobs (any $K = 1, 2, \ldots, n$) must form an optimal schedule for the reduced problem based on just these K jobs alone. To see this, suppose that $(J_{i(1)}, J_{i(2)}, \ldots, J_{i(n)})$ is an optimal schedule for the full problem. Then for any $K = 1, 2, \ldots, n$ we may decompose the performance measure as:

$$
\sum_{k=1}^{n} \gamma_{i(k)}(C_{i(k)}) = \sum_{k=1}^{K} \gamma_{i(k)}(C_{i(k)}) + \sum_{k=K+1}^{n} \gamma_{i(k)}(C_{i(k)})
$$
$$
= \quad A \quad + \quad B \qquad \text{(say)}. \tag{6.1}
$$

Now consider the set of jobs $\{J_{i(1)}, J_{i(2)}, \ldots, J_{i(K)}\}$. Suppose that we face the task of scheduling just these K jobs and not all n. Then surely we cannot improve upon the sequence $(J_{i(1)}, J_{i(2)}, \ldots, J_{i(K)})$. If we could we would be able to reduce the term A in (6.1). Thus we could construct a new sequence for the full problem by using the improved sequence for the first K jobs and leaving the remaining $(n - K)$ in their original order. The total cost of this new sequence would be the sum of a quantity strictly less than A plus the original term B. But this contradicts our basic assumption that $(J_{i(1)}, J_{i(2)}, \ldots, J_{i(n)})$ is an optimal sequence for the full problem. Thus $(J_{i(1)}, J_{i(2)}, \ldots, J_{i(K)})$ must be optimal for the reduced problem.

Before we can use this observation to find an optimal sequence, we need to introduce some notation. We have already been using the first part of this informally. When we write $\{J_1, J_2, \ldots, J_n\}$, we simply mean the set of jobs listed between the curly parentheses; there is no implication of any particular processing order. However, when we use curved brackets, viz. $(J_1, J_2, \ldots, J_n)$, we do wish to imply an order: namely, J_1 first, J_2 second, $\ldots$, and J_n last. In short, the former notation refers to a *set*, the latter to a *sequence*.

If Q is any set of jobs containing the job J, then $Q - \{J\}$ is the set of jobs obtained by deleting J from Q. Also for the set Q we define C_Q by

$$
C_Q = \sum_{J_i \text{ in } Q} p_i.
$$

Thus C_Q is the sum of the processing times for all the jobs in Q. Now remember that by Theorem 3.1 we need not consider inserted idle time when solving an $n/1//B$ problem with B regular. Here we are optimising against the regular measure $\Sigma \gamma_i$; so, when scheduling just the jobs in Q, C_Q is the completion time of the last job processed, whatever sequence is used.

Next we define $\Gamma(Q)$ to be the minimum cost obtained by scheduling the jobs in Q optimally. If Q contains a single job, say $Q = \{J_i\}$, then

$$\Gamma(Q) = \Gamma(\{J_i\}) = \gamma_i(p_i), \tag{6.2}$$

since there is only one way to schedule one job and that job must complete at $C_i = p_i$. Suppose now that Q contains $K > 1$ jobs. Then, remembering that the last job must complete at C_Q and that in any optimal sequence the first $(K - 1)$ jobs are scheduled optimally for a reduced problem involving just those, we have

$$\Gamma(Q) = \min_{J_i \text{ in } Q}\{\Gamma(Q - \{J_i\}) + \gamma_i(C_Q)\}. \tag{6.3}$$

In other words, to find the minimum cost of scheduling Q we consider each job in turn and ask how much it would cost to schedule that job last. We find our answer by taking the minimum of all these possibilities.

Now (6.2) defines $\Gamma(Q)$ for all Q containing a single job. Using these values and (6.3) we can find $\Gamma(Q)$ for all Q containing just two jobs; then for all Q containing just three jobs; and so on until we eventually find $\Gamma(\{J_1, J_2, \ldots, J_n\})$. In doing so we also find an optimal schedule. Perhaps it is easiest to see all this in the context of a particular example.

Example. Solve the $4/1//\bar{T}$ problem with data:

Job J_i	J_1	J_2	J_3	J_4
Processing time, p_i	8	6	10	7
Due date, d_i	14	9	16	16

Here $\gamma_i(C_i) = \frac{1}{4}\max\{C_i - d_i, 0\}$.

First, we calculate $\Gamma(Q)$ for the four sets which only have a single member. For instance by (6.2) we have

$$\Gamma(\{J_1\}) = \gamma_1(C_1) = \tfrac{1}{4}\max\{0, C_1 - d_1\} = \tfrac{1}{4}\max\{0, p_1 - d_1\}$$

Thus we generate:

Table 6.1—$\Gamma(Q)$ for the four single-
job sets

Q	$\{J_1\}$	$\{J_2\}$	$\{J_3\}$	$\{J_4\}$
$p_i - d_i$	−6	−3	−6	−9
$\Gamma(Q)$	0	0	0	0

Next we calculate $\Gamma(Q)$ for the six sets containing just two jobs. For instance, if $Q = \{J_1, J_2\}$, we have $C_Q = p_1 + p_2 = 14$ and by (6.3)

$$\begin{aligned}
\Gamma(Q) &= \min\{\Gamma(\{J_1\}) + \gamma_2(14), \Gamma(\{J_2\}) + \gamma_1(14)\}\\
&= \min\{\quad 0 \quad + \quad 5/4, \quad 0 \quad + \quad 0 \quad\}\\
&= \quad 0.
\end{aligned}$$

The calculations for the five other two-member subsets are laid out in Table 6.2.

Table 6.2—The calculation of $\Gamma(Q)$ for each two-job set

Q	$\{J_1, J_2\}$		$\{J_1, J_3\}$		$\{J_1, J_4\}$		$\{J_2, J_3\}$		$\{J_2, J_4\}$		$\{J_3, J_4\}$	
C_Q	14		18		15		16		13		17	
J_i, 'last' job in sequence	J_1	J_2	J_1	J_3	J_1	J_4	J_2	J_3	J_2	J_4	J_3	J_4
$\gamma_i(C_Q)$	0	$\frac{5}{4}$	1	$\frac{1}{2}$	$\frac{1}{4}$	0	$\frac{7}{4}$	0	1	0	$\frac{1}{4}$	$\frac{1}{4}$
$\Gamma(Q - \{J_i\}) + \gamma_i(C_Q)$	0	$\frac{5}{4}$	1	$\frac{1}{2}$	$\frac{1}{4}$	0	$\frac{7}{4}$	0	1	0	$\frac{1}{4}$	$\frac{1}{4}$
Minimum	*			*		*		*		*	*	
$\Gamma(Q)$	0		$\frac{1}{2}$		0		0		0		$\frac{1}{4}$	

There are two possibilities for each Q corresponding to which job is scheduled last: the columns are divided accordingly. We mark the column corresponding to the minimum in (6.3) by an asterisk, choosing arbitrarily whenever necessary.

We now calculate $\Gamma(Q)$ for the four sets containing three jobs. For instance, if $Q = \{J_1, J_2, J_3\}$, we have $C_Q = p_1 + p_2 + p_3 = 24$ and by (6.3)

$$\begin{aligned}
\Gamma(Q) &= \min\{\Gamma(\{J_1, J_2\}) + \gamma_3(24), \Gamma(\{J_1, J_3\}) + \gamma_2(24),\\
&\qquad\qquad \Gamma(\{J_2, J_3\}) + \gamma_1(24)\}\\
&= \min\{0 + 2, \; 1/2 + 15/4, \quad 0 + 10/4\}\\
&= \quad 2.
\end{aligned}$$

The calculations for the other three-member sets are laid out in Table 6.3. The construction of this exactly parallels that of Table 6.2 except for each Q there are now three possible last jobs and so each column subdivides into three.

Table 6.3—The calculation of $\Gamma(Q)$ for each three-job set

Q	{J₁ J₂, J₃}			{J₁, J₂, J₄}			{J₁, J₃, J₄}			{J₂, J₃, J₄}		
C_Q	24			21			25			23		
J_i, 'last' job in sequence	J_1	J_2	J_3	J_1	J_2	J_4	J_1	J_3	J_4	J_2	J_3	J_4
$\gamma_i(C_Q)$	$\frac{10}{4}$	$\frac{15}{4}$	2	$\frac{7}{4}$	3	$\frac{5}{4}$	$\frac{11}{4}$	$\frac{9}{4}$	$\frac{9}{4}$	$\frac{14}{4}$	$\frac{7}{4}$	$\frac{7}{4}$
$\Gamma(Q-\{J_i\})+\gamma_i(C_Q)$	$\frac{10}{4}$	$\frac{17}{4}$	2	$\frac{7}{4}$	3	$\frac{5}{4}$	3	$\frac{9}{4}$	$\frac{11}{4}$	$\frac{15}{4}$	$\frac{7}{4}$	$\frac{7}{4}$
Minimum			*			*	*					*
$\Gamma(Q)$	2			$\frac{5}{4}$			$\frac{9}{4}$			$\frac{7}{4}$		

Finally we calculate $\Gamma(Q)$ for the set containing all four jobs, viz. $Q = \{J_1, J_2, J_3, J_4\}$. Here we have $C_Q = p_1 + p_2 + p_3 + p_4 = 31$. We calculate $\Gamma(Q)$ in Table 6.4. The construction of this exactly parallels the two previous tables.

Table 6.4—The calculation of $\Gamma(Q)$ for the entire set of four jobs

Q	{J₁, J₂, J₃, J₄}			
C_Q	31			
J_i, 'last' job in sequence	J_1	J_2	J_3	J_4
$\gamma_i(C_Q)$	$\frac{17}{4}$	$\frac{22}{4}$	$\frac{15}{4}$	$\frac{15}{4}$
$\Gamma(Q-\{J_i\})+\gamma_i(C_Q)$	6	$\frac{31}{4}$	5	$\frac{23}{4}$
Minimum			*	
$\Gamma(Q)$	5			

Thus we have found that scheduling the four jobs optimally gives a minimum mean tardiness of 5. Moreover, it is an easy matter to find the optimal schedule. The asterisk in the third column of the Table 6.4 tells us that the minimum mean tardiness is obtained when J_3 is scheduled last. This means that $\{J_1, J_2, J_4\}$ are the first three jobs processed. So we look at Table 6.3 and find that of these J_4 should be scheduled last. Thus we know that $\{J_1, J_2\}$ are the first two jobs processed. Looking at Table 6.2 we find that J_1 should be processed last. This leaves J_2 as the job first processed. So an optimal schedule is (J_2, J_1, J_4, J_3).

This example has illustrated the basic principles of dynamic programming, but there are some subtleties that so far have gone unremarked.

The justification of the dynamic programming approach depended on our being able to decompose the performance measure as in (6.1), repeated here:

$$\sum_{k=1}^{n} \gamma_{i(k)}(C_{i(k)}) = \sum_{k=1}^{K} \gamma_{i(k)}(C_{i(k)}) + \sum_{k=K+1}^{n} \gamma_{i(k)}(C_{i(k)})$$

$$= \quad A \quad + \quad B \quad \text{(say)}.$$

(6.1)

We noted that in any optimal sequence $(J_{i(1)}, J_{i(2)}, \ldots, J_{i(n)})$ the subsequence $(J_{i(1)}, J_{i(2)}, \ldots, J_{i(K)})$ must be optimal for the set of jobs $\{J_{i(1)}, \ldots, J_{i(K)}\}$; otherwise it would be possible to reduce A and hence the sum $(A + B)$.

Now look at this argument more closely. Implicit in it is the assumption that rescheduling the jobs $\{J_{i(1)}, J_{i(2)}, \ldots, J_{i(K)}\}$ cannot affect the term $B = \sum_{k=K+1}^{n} \gamma_{i(k)}(C_{i(k)})$. When we consider semi-active timetabling and zero ready times, this is true. However we reorder jobs $\{J_{i(1)}, J_{i(2)}, \ldots, J_{i(K)}\}$, they complete at time $\sum_{k=1}^{K} p_{i(k)}$. Thus the completion times $C_{i(k)}$ of the jobs in the final sequence $(J_{i(K+1)}, J_{i(K+2)}, \ldots, J_{i(n)})$ are independent of the sequence we use for jobs $\{J_{i(1)}, J_{i(2)}, \ldots, J_{(K)}\}$. In short, we can minimise term A independently of term B. If this independence criterion is not fulfilled, the dynamic programming approach fails. To emphasise this, see Problem 6.6.5. In that problem Held and Karp's algorithm is modified in an attempt to allow for non-zero ready times. But the introduction of non-zero ready times destroys the independence of A and B, and a simple example shows that the modified method may fail to find the optimal solution.

Look again at the decomposition (6.1). From the independence condition we know that an optimal schedule minimises the term A. Hence the first K jobs are also in the right sequence to solve the reduced problem based upon these K jobs alone. Equally we could argue that an optimal schedule must minimise the term B. So we may also deduce that the last $(n - K)$ jobs in an optimal schedule are also in the right sequence to solve the reduced problem based upon these $(n - K)$ jobs alone. But here we must be careful. The term B is minimised subject to the condition that none of the last $(n - K)$ jobs may start before $\sum_{k=1}^{K} p_{i(k)}$, i.e. before all the first K jobs have finished. Thus, to be precise, we should say that the last $(n - K)$ jobs also solve the reduced problem subject to this extra condition. Nonetheless, we may use this observation to develop an alternative dynamic programming solution to the $n/1//\Sigma\gamma_i$ problem.

For any set of jobs Q define

$$s_Q = \sum_{\substack{J_i \text{ not} \\ \text{in } Q}} p_i.$$

Thus s_Q is the earliest that any job in Q can start if all the jobs *not* in Q must be processed first. Next define $\Delta(Q)$ to be the minimum cost incurred from scheduling the jobs in Q subject to the condition that none may start before s_Q. Note that there is no contribution to $\Delta(Q)$ from the jobs not in Q. For any single-job sets we have

$$\Delta(Q) = \Delta(\{J_i\}) = \gamma(s_{\{J_i\}} + p_i),$$

since the single job J_i must complete at $(s_{\{J_i\}} + p_i)$. Noting that $(s_{\{J_i\}} + p_i) = \sum_{j=1}^{n} p_j$, we have

$$\Delta(\{J_i\}) = \gamma_i \left(\sum_{j=1}^{n} p_j \right) \tag{6.4}$$

Generally we find $\Delta(Q)$ by considering each job in Q and calculating the cost of sequencing this job first. $\Delta(Q)$ is the minimum of these costs, viz.

$$\Delta(Q) = \min_{J_i \text{ in } Q} \{\Delta(Q - \{J_i\}) + \gamma_i(s_Q + p_i)\}. \tag{6.5}$$

Using (6.4) and (6.5) we may build up the values of $\Delta(Q)$ for all sets Q. First (6.4) gives $\Delta(Q)$ for all single-job sets; then using (6.5) we obtain $\Delta(Q)$ for all two-job sets; then for all three-job sets; and so on. The procedure is entirely analogous to the manner in which we built up $\Gamma(Q)$.

For an $n/1//\Sigma\gamma_i$ problem it is more natural, perhaps, to base our solution upon expressions (6.2) and (6.3) for $\Gamma(Q)$. Doing so means that we work forward through schedules for the full problem considering subsequences of jobs processed first. For this reason our approach is called **forward dynamic programming.** However, for other problems it is often more natural to base our solution on expressions analogous to (6.4) and (6.5) and consider subsequences of jobs processed last. Such an approach is called **backward dynamic programming.**

6.3 COMPUTATIONAL ASPECTS OF DYNAMIC PROGRAMMING

It is apparent that our dynamic programming solution to the $4/1//\bar{T}$ example of the last section required a lot of calculation, far more than any other solution method that we have met. Admittedly, each step involves only simple additions, subtractions, and comparisons; but the number of them does pose the question: is it all worthwhile. Perhaps it would have been simpler and quicker to have used complete enumeration, i.e. to have listed every possible schedule, then calculated $\bar{T}$ for each, and lastly picked one with the smallest $\bar{T}$. So let us consider the computational merits of complete enumeration on the one hand and dynamic programming on the other.

We shall count the number of mathematical operations that a computer would have to perform in order to solve an $n/1//\bar{T}$ problem by each method. By a mathematical operation is meant the addition of two numbers, or a subtraction, or a comparison. We shall not need multiplications or divisions, but they too are mathematical operations. Very roughly the time a computer takes to solve a problem is directly proportional to the number of mathematical operations that it must perform. (Computer scientists, please forgive our naivety here.) Thus the better method of complete enumeration and dynamic programming is the one that involves the lesser number of such operations.

Suppose that we try to solve an $n/1//\bar{T}$ problem by complete enumeration. There are $n!$ possible permutations of $\{J_1, J_2, \ldots, J_n\}$ and hence $n!$ possible schedules. For each of these we must calculate $\bar{T}$ or, rather $n\bar{T}$, since to find a minimum of the set of mean tardinesses there is no need to divide by the common factor n.

To calculate the total tardiness, $n\bar{T}$, for any schedule we must use an algorithm something like the following:

Algorithm 6.1

Step 1. Set $k = 0$, $C_{i(0)} = 0$, $\Sigma = 0$.
Step 2. Increment k by 1.
Step 3. $C_{i(k)} = C_{i(k-1)} + p_{i(k)}$; $L_{i(k)} = C_{i(k)} - d_{i(k)}$; $T_{i(k)} = \max\{0, L_{i(k)}\}$
Step 4. Add $T_{i(k)}$ to Σ
Step 5. Is $k = n$? If so, stop. If not, go to *Step 2*.

For the schedule $(J_{i(1)}, J_{i(2)}, \ldots, J_{i(n)})$ this algorithm cycles through the jobs in their processing order. For each job it calculates the completion time, the lateness, and the tardiness, finally adding the tardiness into the rolling total Σ. Thus at the completion of the cycle Σ is the total tardiness of the schedule. Clearly, this algorithm involves the following operations.

n	additions to increment the index k through $k = 1, 2, \ldots, n$.
n	additions to calculate the completion times.
n	subtractions to calculate the lateness.
n	comparisons to calculate the tardiness.
n	additions to accumulate the rolling total Σ.
n	comparisons to determine whether $k = n$.
$\overline{6n}$	

Thus $6n$ operations are required to calculate the total tardiness for each of the $n!$ possible schedules.

To find the minimum total tardiness we must use an algorithm something like the one below. I assume that the set of all $n!$ possible schedules is indexed by j.

Algorithm 6.2

Step 1. Set $\Sigma_{\min} = \Sigma_1$, $\mu = 1$, $j = 1$.
Step 2. Increment j by 1.
Step 3. Is $\Sigma_{\min} \leqslant \Sigma_j$? If so, go to *Step 5.*
Step 4. Reset $\Sigma_{\min}$ to Σ_j and reset μ to j.
Step 5. Is $j = (n!)$? If so, stop. If not, go to *Step 2.*

Steps 3 and 4 ensure that $\Sigma_{\min}$ is always equal to the smallest total tardiness encountered so far. At the end of this algorithm μ is the index of an optimal schedule and $\Sigma_{\min}$ is the minimum total tardiness attained. The above involves:

$(n! - 1)$ additions to increment the index j through $j = 1, 2, \ldots, n!$;
$(n! - 1)$ comparisons to locate the minimum total tardiness;
$\underline{(n! - 1)}$ comparisons to determine whether $j = n!$
$3(n! - 1)$

Thus complete enumeration requires

$$6n(n!) + 3(n! - 1) \tag{6.6}$$

mathematical operations to solve an $n/1//\bar{T}$ problem.

Next consider the dynamic programming approach. Here we shall need the notation:

$$\binom{n}{K} = \frac{n!}{K!(n - K)!}$$

= number of distinct subsets of exactly K jobs which may be formed from the total set of n jobs.

If we say that a subset has size K when it contains K jobs, then we see that dynamic programming calculates $\Gamma(Q)$ for

$\binom{n}{1}$ subsets Q of size 1,

$\binom{n}{2}$ subsets Q of size 2,

$\binom{n}{3}$ subsets Q of size 3,

. . . .
. . . .

and $\binom{n}{n}$ subsets Q of size n.

We shall not explicitly write out the algorithms needed to perform the calculations for each subset Q. Instead we note that to control a loop around which the calculation cycles N times requires $2N$ operations, 1 increment and 1 comparison for each cycle. (Check this in the algorithms above.) Now for any subset Q of size K we first calculate C_Q, then we cycle K times considering in turn the possibility of scheduling each job in Q last. (Look back to the example of the last section and, in particular, the construction of the tables.) Thus the following mathematical operations must be performed.

  1 addition to find C_Q
     (N.B. only 1 addition is needed; see the example of Section 3.4)
  K subtractions to find the lateness of each possible last job, $L_i = C_Q - d_i$.
  K comparisons to find the tardiness of each possible last job, $T_i = \max\{L_i, 0\}$.
  K additions to form each sum: $\Gamma(Q - \{J_i\}) + \gamma_i(C_Q)$.
  K comparisons to find the minimum of these sums, $\Gamma(Q)$.
  $2K$ operations to control the loop which considers in turn the possibility of each job in Q being scheduled last.
  ─────
  $6K + 1$

Apart from counting these operations, we need to remember that there is another loop which takes us from one subset Q to the next, starting with subsets of size 1, cycling through all these, then moving to subsets of size two, and so on. This can be done by a loop which involves 2 operations for each subset Q (see e.g. Baker and Schrage, 1978). Thus the dynamic programming method requires $(6K + 3)$ operations per subset Q of size K. In total, therefore, it requires:

$$\binom{n}{1}(6.1 + 3) + \binom{n}{2}(6.2 + 3) + \binom{n}{3}(6.3 + 3) + \ldots + \binom{n}{n}(6n + 3)$$

operations. It can be shown (see Problem 6.6.8) that this sum is:

$$6n.2^{n-1} + 3(2^n - 1). \tag{6.7}$$

We are now in a position to compare the number of operations required by complete enumeration as opposed to the number required by dynamic programming. Table 6.5 gives the values of expressions (6.6) and (6.7) for various values of n.

Most people find the numbers in Table 6.5 quite startling when seeing them for the first time. If a computer takes 1 microsecond to perform one operation (1 microsecond $= 10^{-6}$ second), then complete enumeration will take ten million years to solve a 20/1/T problem, whereas dynamic programming will solve the problem in about 1 minute. Why is dynamic programming so much faster?

Table 6.5—Comparison of the computation required to solve an $n/1//\overline{T}$ problem by complete enumeration and by dynamic programming

	Number of operations required by	
n	Complete enumeration	Dynamic Programming
4	647	237
10	2.286×10^8	33789
20	2.992×10^{20}	6.396×10^7
40	1.983×10^{50}	1.352×10^{14}

Essentially complete enumeration first lists all the possible schedules, and only when it has done this does it eliminate any. Dynamic programming eliminates many possible schedules as it constructs the list. Thus in the example of the last section the method had eliminated by the second stage (Table 6.2) all processing sequences beginning: $(J_1, J_2, -, -)$, $(J_3, J_1, -, -)$, $(J_4, J_1, -, -)$ $(J_3, J_2, -, -)$, $(J_4, J_2, -, -)$, and $(J_3, J_4, -, -)$. These sequences correspond to the unasterisked columns in the table. Since there are two possible sequences with each of these beginnings, dynamic programming has eliminated 12 non-optimal processing sequences at this early stage of the calculation. There are only 24 possible processing sequences in the problem, so the method is gaining considerably over complete enumeration.

That dynamic programming is to be preferred to complete enumeration as a method of solving these problems is clear. It is faster. Moreover, the larger the problem the more dramatic is the gain in speed. However, before we get too overjoyed at finding a method of solving our problems, let us stop and think. To solve a $20/1//\overline{T}$ problem dynamic programming may take about a minute, if we assume the computational speed of 1 operation per microsecond; but a $40/1//\overline{T}$ problem will take over 4 years, far too long for most computer budgets. So dynamic programming only gives us a way of solving problems with up to about 25 jobs.

There is a further, more difficult problem associated with this dynamic programming method. It has to store and remember very many intermediate calculations. For instance, it has to remember for each subset Q which job to schedule last; i.e. it has to remember where the asterisks are in the tables. For only when all the $\Gamma(Q)$ have been calculated, is it possible to step back through the tables picking out the appropriate asterisks and so finding an optimal schedule. There are 2^n asterisks to be remembered (why?). Computers have finite memories. (They can only remember a certain number of things.) For small computers the limit is about 2^{12} and for larger ones it may stretch up to 2^{20}. So this puts a further limit on the size of problem that dynamic programming can be used to solve.

(N.B. I am making an implicit distinction here between 'core' storage and 'magnetic disc' storage. If a magnetic disc is used, a computer can store very many more quantities than 2^{20}. However, reading them back from disc is very slow compared with recalling them from core. Thus using magnetic disc helps solve the problem of storage, but adds to the problem of time.)

In conclusion, we see that the dynamic programming approach enables us to solve small $n/1//\Sigma\gamma_i$ problems, but it does not offer us a way of tackling larger ones. However, perhaps all is not lost. Consider what happens when we introduce precedence constraints.

6.4 DYNAMIC PROGRAMMING SUBJECT TO PRECEDENCE CON-STRAINTS

In section 4.1 I made the rather remarkable statement that the introduction of precedence constraints into a problem, far from complicating matters, can actually make it easier to solve. Let me now show you that this is indeed so.

Consider $5/1//\overline{T}$ problem with data:

Job	J_1	J_2	J_3	J_4	J_5
Processing time, p_i	4	2	6	3	5
Due date, d_i	7	9	9	6	12

Assume, furthermore, that precedence constraints demand that J_1 is completed before either J_2 or J_3 is started and that J_4 is completed before J_5 is started. These constraints are shown in Fig. 6.1. Note that there is no demand that, say, J_4 immediately precedes J_5; other jobs may intervene.

Fig. 6.1 Precedence constraints for the example.

So let us begin to calculate $\Gamma(Q)$ for each possible set of jobs. Beginning with the single job-sets, we see immediately that there is no point in considering the three sets $\{J_2\}, \{J_3\}$, and $\{J_4\}$ because none of these can be the first job in a schedule. Thus our first table has two columns instead of five.

Table 6.6—$\Gamma(Q)$ for the two possible single-job sets

Q	$\{J_1\}$	$\{J_4\}$
$p_i - d_i$	-3	-3
$\Gamma(Q)$	0	0

Next we consider the two-job sets. Given that there are five jobs in all, there should be $\binom{5}{2} = 10$ sets to consider here. However, only four of these are compatible with the procedence constraints; for instance, $\{J_1, J_5\}$ cannot be the first two jobs in any schedule, because J_4 must precede J_5. Moreover, when we consider those two-job sets that are possible, we find that we have no choice about which to sequence last in three of the four cases; the precedence constraints dictate the order. Thus, in Table 6.7 instead of ten columns each subdividing into two we have four columns only one of which subdivides.

Table 6.7—The calculation of $\Gamma(Q)$ for the four possible two-job sets

Q	$\{J_1, J_2\}$		$\{J_1, J_3\}$		$\{J_1, J_4\}$		$\{J_4, J_5\}$	
C_Q	6		10		7		8	
J_i, 'last' job in sequence	J_1	J_2	J_1	J_3	J_1	J_4	J_4	J_5
$\gamma_i(C_Q)$		0		$\frac{1}{5}$	0	$\frac{1}{5}$		0
$\Gamma(Q - \{J_i\}) + \gamma_i(C_Q)$		0		$\frac{1}{5}$	0	$\frac{1}{5}$		0
Minimum	Impossible	*	Impossible	*	*		Impossible	*
$\Gamma(Q)$	0		$\frac{1}{5}$		0		0	

As we continue we find similar reduction in the number of calculations that we have to perform.

Table 6.8—The calculation of $\Gamma(Q)$ for the four possible three-job sets

Q	$\{J_1, J_2, J_3\}$			$\{J_1, J_2, J_4\}$			$\{J_1, J_3, J_4\}$			$\{J_1, J_4, J_5\}$		
C_Q	12			9			13			12		
J_i, 'last' job in sequence	J_1	J_2	J_3	J_1	J_2	J_4	J_1	J_3	J_4	J_1	J_4	J_5
$\gamma_i(C_Q)$	Impossible	$\frac{3}{5}$	$\frac{3}{5}$	Impossible	0	$\frac{3}{5}$	Impossible	$\frac{4}{5}$	$\frac{7}{5}$	$\frac{5}{5}$	Impossible	0
$\Gamma(Q - \{J_i\}) + \gamma_i(C_Q)$		$\frac{4}{5}$	$\frac{3}{5}$		0	$\frac{3}{5}$		$\frac{4}{5}$	$\frac{8}{5}$	$\frac{5}{5}$		0
Minimum			$*$		$*$			$*$				$*$
$\Gamma(Q)$	$\frac{3}{5}$			0			$\frac{4}{5}$			0		

Table 6.9—The calculation of $\Gamma(Q)$ for the three possible four-job sets

Q	$\{J_1, J_2, J_3, J_4\}$				$\{J_1, J_2, J_4, J_5\}$				$\{J_1, J_3, J_4, J_5\}$			
C_Q	15				14				18			
J_i, 'last' job in sequence	J_1	J_2	J_3	J_4	J_1	J_2	J_4	J_5	J_1	J_3	J_4	J_5
$\gamma_i(C_Q)$	Impossible	$\frac{6}{5}$	$\frac{6}{5}$	$\frac{9}{5}$	Impossible	$\frac{5}{5}$	Impossible	$\frac{2}{5}$	Impossible	$\frac{9}{5}$	Impossible	$\frac{6}{5}$
$\Gamma(Q - \{J_i\}) + \gamma_i(C_Q)$		$\frac{10}{5}$	$\frac{6}{5}$	$\frac{12}{5}$		$\frac{5}{5}$		$\frac{2}{5}$		$\frac{9}{5}$		$\frac{10}{5}$
Minimum				$*$				$*$			$*$	
$\Gamma(Q)$	$\frac{6}{5}$				$\frac{2}{5}$				$\frac{9}{5}$			

Thus stepping back through the tables picking out the appropriate asterisks, we find that the optimal schedule is $(J_4, J_1, J_2, J_5, J_3)$. The mean tardiness for this schedule is 13/5.

The solution of this $5/1//\bar{T}$ problem required no more effort than our solution of the $4/1//\bar{T}$ problem in Section 6.2. Yet, had there been no precedence constraints, it would have required much more. Indeed, using expression (6.7) we find that the solution of a 5 job problem without precedence constraints requires about two and a half times as much work as that of a 4 job one. So we see that the introduction of precedence

Table 6.10—The calculation of $\Gamma(Q)$ for the entire set of five jobs

Q	$\{J_1, J_2, J_3, J_4, J_5\}$				
C_Q	20				
J_i, 'last' job in sequence	J_1	J_2	J_3	J_4	J_5
$\gamma_i(C_Q)$		$\frac{11}{5}$	$\frac{11}{5}$		$\frac{8}{5}$
$\Gamma(Q - \{J_i\}) + \gamma_i(C_Q)$	Impossible	$\frac{20}{5}$	$\frac{13}{5}$	Impossible	$\frac{14}{5}$
Minimum			*		
$\Gamma(Q)$	$\frac{13}{5}$				

constraints makes our solution of $n/1//\Sigma\gamma_i$ problems by dynamic programming easier. Exactly how much easier depends on the particular precedence constraints, but clearly a substantial reduction in difficulty is possible. Since very many practically occurring problems involve precedence constraints, our conclusions in the previous section may seem over-pessimistic. Moreover, there may yet be a way to make our solution of problems without precedence constraints easier. Consider the following theorem.

Theorem 6.3. In an $n/1//\bar{T}$ problem if two jobs J_i and J_k are such that (i) $p_i \leq p_k$ and (ii) $d_i \leq d_k$, then there exists an optimal schedule in which J_i precedes J_k.

Proof. See Problem 6.6.10.

This theorem is typical of a class of results called **dominance conditions or elimination criteria.** These take the form: if J_i is related to J_k in a particular way, then there exists an optimal schedule in which J_i precedes J_k. Suppose that we have an $n/1//\bar{T}$ problem and suppose that we find a dominance condition applies to the pair of jobs J_i and J_k. Then we know that an optimal schedule exists in which J_i precedes J_k. Hence we may introduce the precedence constraint that J_i should precede J_k and be sure that a solution to the modified problem also solves the original one. Moreover, the introduction of the precedence constraint makes the problem easier to solve by dynamic programming. Thus if we examine enough pairs of jobs and find enough dominance conditions holding, we might introduce enough precedence constraints to make quite large problems easily solvable by dynamic programming.

Empirical investigations have shown that this use of dominance conditions to introduce precedence constraints into a problem can lead to very fast dynamic programming algorithms for solving single machine problems (Baker and Schrage, 1978; Schrage and Baker, 1978; and Van Wassenhove and Gelders, 1978). However, there is a difficulty, which limits the applicability of the algorithm; it makes substantial demands on the computer's memory. As we noted in the previous section, there are many intermediate quantities which must be remembered throughout the calculation. Even with the reduction in their number brought by the introduction of precedence constraints, there are still sufficiently many to make problems with more than about 25 jobs intractable.

6.5 REFERENCES AND FURTHER READING

The material of Section 6.2 was taken from Held and Karp (1962). Lawler (1964) and Baker (1974) present the backward dynamic programming formulation of the same problem. Lawler and Moore (1969) develop some further applications of dynamic programming to simply structured scheduling problems. Conway, Maxwell, and Miller (1967) and Baker (1974) both discuss the solution of problems with sequence dependent set up times. See also Corwin and Esogbue (1974).

Baker and Schrage's (1978) paper was the first to extend the use of dynamic programming to solve problems in which precedence relations occur. Their study mainly concerned required strings of jobs of the form discussed in Section 4.1. However, they were the first to suggest the use of dominance relations to introduce precedence constraints into problems where *a priori* there are none. This suggestion has been investigated by Schrage and Baker (1978), and Gelders and Van Wassenhove (1978). The dominance condition that we quoted (Theorem 6.3) may be found along with some more general conditions in Rinnooy Kan *et al.* (1975), and Rinnooy Kan (1976, pp 71–76). Both these works warn of the danger of introducing precedence cycles if too many dominance conditions are applied simultaneously. Their remarks and their method of avoiding the difficulty are equally applicable to the use of such conditions in dynamic programming. See also Smeds (1980).

Translating the 'pencil and paper' approach to dynamic programming to a computer is by no means straightforward, particularly when precedence constraints are involved. Baker and Schrage (1978) suggest one way of doing this, but Lawler (1979) argues that their way is not the most efficient and proposes an alternative.

6.6 PROBLEMS

1. Solve the $4/1//\bar{T}$ problem with data:

Job J_i	J_1	J_2	J_3	J_4
Processing time	9	12	7	14
Due date	15	19	23	21

2. Solve the $4/1//\Sigma\gamma_i(C_i)$ problem for the data given in Problem 6.6.1 where $\Sigma_{i=1}^{n}\gamma_i(C_i) = 6T_1 + 2T_2 + T_3 + 9T_4$.

3. Consider applying dynamic programming to solve the general $n/1//\bar{F}$ problem and so deduce that an SPT schedule provides an optimal solution.

4. Solve the example of Section 6.2 by backward dynamic programming.

5. Throughout this chapter we have assumed that the ready times of all jobs are zero. Suppose we tried to modify our dynamic programming approach to treat problems where this is not the case. If Q contains a single job, $Q = \{J_i\}$, then define

$$\Gamma(Q) = \Gamma(\{J_i\}) = \gamma_i(r_i + p_i)$$

$$C(Q) = C(\{J_i\}) = r_i + p_i.$$

If Q contains more than one job, then for J_i in Q let

$$t_i = \max\{r_i, C(Q - \{J_i\})\} + p_i,$$

and define

$$\Gamma(Q) = \min_{J_i \text{ in } Q}\{\Gamma(Q - \{J_i\}) + \gamma_i(t_i)\};$$

$$C(Q) = t_{i^*}, \quad \text{where } J_{i^*} \text{ is the job that produces}$$
$$\text{the minimum in } \Gamma(Q) \text{ above.}$$

Essentially this modification tries to allow for the possibility that the 'last' job J_i in the set Q may not be ready to start immediately the jobs in $Q - \{J_i\}$ have finished. Show that this algorithm does *not* work by considering the $3/1//\Sigma\gamma_i(C_i)$ example below.

(*Hint:* (J_1, J_2, J_3) is optimal).

Job	J_1	J_2	J_3
Processing time	5	5	5
Ready time	0	1	10
Due date	11	9	15
$\gamma_i(C_i)$	T_1	T_2	$3T_3$

6. Discuss how you would extend the dynamic programming method presented here to deal with performance measures of the form $\max_{i=1}\{\gamma_i(C_i)\}$. Derive the necessary recurrence equations. Use your method to solve the $4/1//L_{max}$ problem with data:

Job	J_1	J_2	J_3	J_4
Processing time	5	8	3	6
Due date	10	9	7	11
Ready time	0	0	0	0

Use Lawler's method to solve the same problem.

N.B. Lawler's method applies equally well when there are no precedence constraints. In both case you will, of course, find that the earliest due date sequence is optimal, but you should note the relative computational merits of your two methods of solution.

7. Consider an $n/1//C_{max}$ problem in which the processing times are *not* sequence independent. In particular, this allows for sequence dependent set-up times, since these are included in the processing times. Let π_{ij} be the processing time of J_j if it immediately follows J_i, and π_{oj} be that of J_j if it is first in the processing sequence. For any set of jobs Q which contains the job J_i let $\Gamma(Q,j)$ be the minimum time required to process all the jobs in Q subject to the condition that J_j is the last processed. Show that for single job sets:

$$\Gamma(\{J_j\},j) = \pi_{oj}$$

and for sets of more than one job:

$$\Gamma(Q,j) = \min_{J_i \text{ in } Q-\{J_j\}} \{\Gamma(Q - \{j\},i) + \pi_{ij}\}.$$

Deduce that the minimum make-span for the $n/1//C_{max}$ problem is

$$\min_{j=1}^{n} \{\Gamma(\{J_1, J_2, \ldots, J_n\},j)\}.$$

Hence solve the $4/1//C_{max}$ problem with sequence dependent processing times given by:

	π_{ij}	j = 1	2	3	4
	0	3	2	6	1
	1	–	5	9	3
i	2	2	–	4	1
	3	6	5	–	3
	4	8	2	6	–

8. The binomial expansion gives:

$$(1 + x)^n = 1 + \binom{n}{1}x + \binom{n}{2}x^2 + \binom{n}{3}x^3 + \ldots + \binom{n}{n}x^n.$$

Use this to show that

(i) $2^n = 1 + \binom{n}{1} + \binom{n}{2} + \binom{n}{3} + \ldots + \binom{n}{n}$

(ii) $n2^{n-1} = 1.\binom{n}{1} + 2.\binom{n}{2} + 3.\binom{n}{3} + \ldots + n.\binom{n}{n}.$

Use these expressions to derive the formula (6.7) for the number of mathematical operations required by dynamic programming.

9. Solve the $6/1//\bar{T}$ problem with the precedence constraints that J_1 must be processed before both J_2 and J_3, and that J_4 must be processed immediately before J_5, which in turn must be processed immediately before J_6. The processing times and due dates of the jobs are:

J_i	J_1	J_2	J_3	J_4	J_5	J_6
p_i	6	4	8	2	10	3
d_i	9	12	15	8	20	22

10. Prove Theorem 6.3.

11. Apply Theorem 6.3 to the example of Section 6.2, so introducing two precedence constraints. Consider carefully the reduction in computation that these allow.

Chapter 7

Branch and Bound Methods

7.1 INTRODUCTION

Apart from heuristic methods, **branch and bound** is probably the solution technique most widely used in scheduling. Like dynamic programming, it is an enumeration technique and, again like dynamic programming, it is an approach to optimisation which applies to a much larger class of problems than just those that arise in our subject. However, we shall discuss it only in the context of scheduling; Agin (1966), and Lawler and Wood (1966) give more general surveys.

We shall motivate the branch and bound approach by a further study of the logical structure of dynamic programming. This is not to say that the former is an extension of the latter: far from it. Nonetheless it is related, and appreciating that relationship will both further our understanding of dynamic programming and serve as a useful introduction to our study of branch and bound.

Although our examples will be based upon fairly simple classes of problems, there are no specific limitations to the branch and bound method. Thus we make no special assumptions beyond those listed in Chapter 1.

7.2 DYNAMIC PROGRAMMING AND ITS ELIMINATION TREE

We begin by recalling the $4/1//\bar{T}$ example of Section 6.2 and considering how dynamic programming eliminates the non-optimal schedules. A schedule for this problem corresponds to an assignment of a different job to each of the four positions in the processing sequence. We may generate all $4! = 24$ possible schedules through the hierarchial, branching structure shown in Fig. 7.1. Initially ignore the distinction between solid and dotted lines. To generate a schedule we start with no job sequenced and indicate this by the point or **node XXXX**. Throughout, an 'X' in the processing sequence will indicate that no job has yet been assigned that position. We

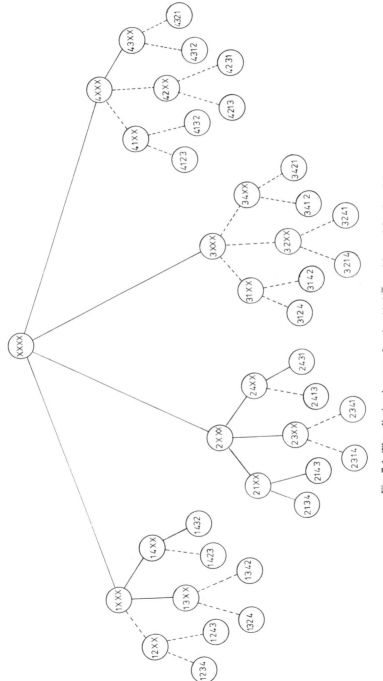

Fig. 7.1 The elimination tree for the $4/1//\overline{T}$ problem of Section 6.2.

begin by assigning a job to the first position in the sequence. Thus we move from node **XXXX** to one of the four mutually exclusive and exhaustive possibilities: 1**XXX**, 2**XXX**, 3**XXX**, and 4**XXX**. Next we assign the second job in the sequence, so branching from each of these four nodes to three possibilities. Thus 1**XXX** branches to the nodes 12**XX**, 13**XX**, and 14**XX**; 2**XXX** branches to the nodes 21**XX**, 23**XX**, and 24**XX**; and so on. Finally we again subdivide our possibilities by assigning the job to the processed third. Since there are only four jobs, this immediately fixes the last job. Thus, for example, 12**XX** branches to the nodes 1234 and 1243.

For reasons that should be apparent, the branching structure in Fig. 7.1 is called a **tree**, more particularly, because we use it to consider the elimination of non-optimal schedules, an **elimination tree**.

Consider the calculation of the mean tardiness of any schedule, say 2413. We start by taking the job scheduled first, calculating its completion time, its tardiness and, hence, its contribution to $\bar{T}$. This calculation can be represented by a move from node **XXXX** to node 2**XXX**. The calculation of the contribution to $\bar{T}$ of the job scheduled second may similarly be represented by a move on from node 2**XXX** to node 24**XX**. Third and fourth we calculate the tardinesses of jobs 1 and 3. We should represent these calculations by moves from 24**XX** to 241**X** and then from 241**X** to 2413. However, nodes 241**X** and 2413 are amalgamated in our diagram so this move is not fully indicated. (Drawing in the extra nodes 241**X** etc. complicates the figure to little purpose.) Hence the calculation of $\bar{T}$ for any schedule may be represented by moves along branches of the tree starting at **XXXX** and travelling to the appropriate final node. Each time a node is encountered, a further contribution to $\bar{T}$ is evaluated.

With this tree representation we may discuss the structure of different solution methods. For example, complete enumeration systematically starts from node **XXXX** twenty four times and travels to each and every final node. In doing so it calculates all the $\bar{T}$'s. Section 6.3 showed us just how many computations this requires.

Dynamic programming again starts at **XXXX**; but here there are four simultaneous moves to nodes 1**XXX**, 2**XXX**, 3**XXX** and 4**XXX**. These correspond to the calculations in Table 6.1, which evaluate $\Gamma(Q)$ for the single-member sets. Next consider the calculation of $\Gamma(Q)$ for the two-member sets. For instance, look at $Q = \{J_1, J_2\}$ in Table 6.2. From our calculations we determine that job 1 should be scheduled last. In terms of the elimination tree, moving from 2**XXX** to 21**XX** is better than moving from 1**XXX** to 12**XX**. This we indicate by a solid line for the former and a dotted for the latter. Moreover, we also know that we need not consider nodes that lie beyond 12**XX** in our tree. So we also use dotted lines to join node 12**XX** to both 1234 and 1243. In short, we have eliminated a pair of final nodes from further consideration. Similarly our calculations for each of the other five two-member sets eliminate five other pairs of final nodes. Thus of the twenty-four possible schedules we have already eliminated twelve.

Turning to the calculation of $\Gamma(Q)$ for three-member sets Q, we see that further schedules are eliminated. Consider $Q = \{J_1, J_2, J_3\}$. Here we learn that J_3 should be scheduled last. Thus we move from 21XX to 2134 and eliminate moves from 13XX to 1324 and from 23XX to 2314. Similarly three other pairs of schedules are eliminated in calculating $\Gamma(Q)$ for the three other sets. So of the twenty-four possible schedules we have now eliminated twenty. The final calculation of $\Gamma(\{J_1, J_2, J_3, J_4\})$ selects an optimal schedule from the remaining four.

Hence we see the advantage that dynamic programming holds over complete enumeration. The latter explores every possible path from XXXX to an extremity, whereas the former eliminates many possible paths on route.

The above not only gives some insight into the advantages of dynamic programming, but also points to some of its problems. The method moves out *simultaneously* from XXXX in all directions. it is always working in many different parts of the tree at the time. For example, it eliminates the move 43XX to 4312 by consideration of the move from 14XX to 1432. This simultaneous working puts extreme demands on the memory requirements. It must remember much information at each and every stage and none of this information becomes redundant until the final selection of the optimal policy. Branch and bound tries to avoid these failings. Firstly, it moves out from XXXX in a very uneven fashion exploring some branches fully before looking at others at all. Secondly, it continually checks to see if some of stored information has become redundant and, if so, this is promptly forgotten.

Because both branch and bound and dynamic programming explore the elimination tree intelligently, determining on route which branches need not be fully investigated, they are called **implicit enumeration** methods. Implicit in their logic is the checking of every possible schedule, but unlike **complete or explicit enumeration** they do not consider every possibility explicitly.

7.3 A FLOW-SHOP EXAMPLE

We shall introduce the underlying ideas of branch and bound through a simple $4/3/F/C_{max}$ example with zero ready times. Thus C_{max} and F_{max} are equivalent performance measures for this example. The specific formulation that we adopt was developed independently by Ignall and Schrage (1965) and by Lomnicki (1965).

Note that by Theorems 5.1 and 5.2 we need only consider permutation schedules. Thus the elimination tree for this problem will have an identical branching structure to that of Fig. 7.1. Central to the technique of branch and bound is, not surprisingly, the idea of bounding. Suppose we are at node 2XXX in the elimination tree. Then we shall calculate a lower bound on C_{max} for the six possible schedules that lie beyond this node, viz. 2134, 2143, 2314, 2341, 2413 and 2431. Consideration of this bound will tell us whether to explore this branch of the tree any further. But that is looking ahead. First we must discuss the bounds themselves; then we can discuss their use.

For notational convenience we follow our convention in Chapter 5 and write

$$a_i = p_{i1}, \, b_i = p_{i2}, \quad \text{and} \quad c_i = p_{i3} \quad \text{for all jobs } J_i.$$

It will not hurt to generalise our problem to n jobs while we develop the form of the bounds. Suppose we are at a node $J_{i(1)}, J_{i(2)}, \ldots, J_{i(K)} XX \ldots X$. Thus K jobs $\{J_{i(1)}, J_{i(2)}, \ldots J_{i(K)}\}$ have been assigned to the first K positions in the processing sequence. We let $A = (J_{i(1)}, J_{i(2)}, \ldots, J_{i(K)})$ be the subsequence of jobs already assigned and U the set of $(n - K)$ jobs as yet unassigned. For $k = 1, 2, \ldots, K$ we let $\alpha_{i(k)}$ be the completion time of job $J_{i(k)}$ on machine 1, $\beta_{i(k)}$ its completion time on machine 2, and $\gamma_{i(k)}$ its completion time on machine 3. $\alpha_{i(k)}$, $\beta_{i(k)}$, and $\gamma_{i(k)}$ may be calculated recursively as:

$$\alpha_{i(1)} = a_{i(1)}; \qquad\qquad \alpha_{i(k)} = \alpha_{i(k-1)} + a_{i(k)};$$
$$\beta_{i(1)} = a_{i(1)} + b_{i(1)}; \qquad \beta_{i(k)} = \max\{\alpha_{i(k)}, \beta_{i(k-1)}\} + b_{i(k)}; \quad (7.1)$$
$$\gamma_{i(1)} = a_{i(1)} + b_{i(1)} + c_{i(1)}; \quad \gamma_{i(k)} = \max\{\beta_{i(k)}, \gamma_{i(k-1)}\} + c_{i(k)}.$$

To develop the lower bound we consider three possibilities that are, in a sense, the most favourable that can occur. We investigate how quickly we may complete the remaining jobs in U if we compact the processing on each of the three machines in turn. First we consider the possibility that the processing on machine 1 is continuous; this is, of course, always true for any semi-active schedule. But suppose that the last job in the schedule, say $J_{i(n)}$, does not have to wait for either machine 2 or machine 3 to be free before being processed on these. Since $C_{max} = C_{i(n)}$, we have

$$C_{max} = \alpha_{i(K)} + \sum_{J_i \text{ in } U} a_i + (b_{i(n)} + c_{i(n)}).$$

Thus, choosing $J_{i(n)}$ to have the least total processing on machines 2 and 3, we see that all schedules must have

$$C_{max} \geq \alpha_{i(K)} + \sum_{J_i \text{ in } U} a_i + \min_{J_i \text{ in } U}\{b_i + c_i\}. \quad (7.2)$$

Next we consider the possibility that the processing on machine 2 is continuous. To ensure this we must assume that there is no idle time while it waits for jobs still being processed on machine 1. Further suppose that the last job in the schedule, $J_{i(n)}$, can complete on machine 3 without needing to wait for $J_{i(n-1)}$ to complete. Then we have

$$C_{max} = \beta_{i(K)} + \sum_{J_i \text{ in } U} b_i + c_{i(n)}.$$

Choosing $J_{i(n)}$ to have the least possible processing time on machine 3, we see that all schedules must have

$$C_{max} \geq \beta_{i(k)} + \sum_{J_i \text{ in } U} b_i + \min_{J_i \text{ in } U}\{c_i\}. \tag{7.3}$$

Finally we suppose jobs process on machines 1 and 2 such that there is no need for idle time on machine 3; i.e. processing on that machine may be continuous. Thus we see that all schedules must have

$$C_{max} \geq \gamma_{i(K)} + \sum_{J_i \text{ in } U} c_i \tag{7.4}$$

Now since no schedule can do better than any of expressions (7.2), (7.3) and (7.4), no schedule can do better than the maximum of these. Thus we have the lower bound:

$$lb(\mathbf{A}) = \max\{\alpha_{i(K)} + \sum_{J_i \text{ in } U} a_i + \min_{J_i \text{ in } U}\{b_i + c_i\},$$
$$\beta_{i(K)} + \sum_{J_i \text{ in } U} b_i + \min_{J_i \text{ in } U}\{c_i\}, \tag{7.5}$$
$$\gamma_{i(K)} + \sum_{J_i \text{ in } U} c_i\}.$$

This lower bound may be strict; i.e. it may be strictly less than the makespan of all schedules beginning with the subsequence A. The reason is simple. In deriving the three components of the lower bound we have assumed that there need not be waiting time nor idle time at certain points in the schedule. In many cases this need not be so.

With these definitions in mind we now turn to the solution of a particular $4/3/F/C_{max}$ problem. The table below gives the processing times. We assume that all the ready times are zero.

Job	Processing Times		
J_i	a_i	b_i	c_i
1	1	8	4
2	2	4	5
3	6	2	8
4	3	9	2

For convenience in calculating $\min_{J(i) \text{ in } U} \{b_i + c_i\}$ and $\min_{J(i) \text{ in } U} \{c_i\}$ for various sets of unassigned jobs U, we also tabulate $(b_i + c_i)$ and c_i in increasing order

J_i	$b_i + c_i$		J_i	c_i
2	9		4	2
3	10		1	4
4	11		2	5
1	12		3	8

We begin by calculating lower bounds on C_{max} at all the nodes on the branch of the elimination tree leading to the processing sequence 1234; i.e. we travel from **XXXX** to 1**XXX** to 12**XX** and finally to 1234. Naturally, at the final node 1234 we calculate C_{max} itself and not a lower bound. Thus we obtain:

At 1XXX
$A = (J_1), U = \{J_2, J_3, J_4\}$.
$\alpha_1 = 1, \beta_1 = 9, \gamma_1 = 13$.
$lb(A) = \max\{1 + 11 + 9, 9 + 15 + 2, 13 + 15\}$
$\qquad = 28.$

At 12XX
$A = (J_1, J_2), U = \{J_3, J_4\}$.
$\alpha_2 = 1 + 2 = 3, \beta_2 = \max\{3,9\} + 4 = 13$.
$\gamma_2 = \max\{13, 13\} + 5 = 18$.
$lb(A) = \max\{3 + 9 + 10, 13 + 11 + 2, 18 + 10\} = 28.$

At 1234
$A = (J_1, J_2, J_3, J_4)$; there are no unassigned jobs.
$\alpha_3 = 3 + 6 = 9, \beta_3 = \max\{9, 13\} + 2 = 15,$
$\gamma_3 = \max\{15, 18\} + 8 = 26.$

$\alpha_4 = 9 + 3 = 12, \beta_4 = \max\{12, 15\} + 9 = 24,$
$\gamma_4 = \max\{24, 26\} + 2 = 28.$

Hence
C_{max} for schedule $1234 = \gamma_4 = 28.$

Thus we find the C_{max} for the schedule 1234 is 28. 1234 is our first **trial schedule**. Our scheme will be to explore the elimination tree comparing the lower bounds at each node with the value of C_{max} for the current trial schedule. If the lower bound at a node is greater than or equal to this we know that we cannot improve upon the trial schedule by exploring that branch further, and hence we eliminate that node and all nodes beyond it in the branch. If the lower bound at a node is less than C_{max} of the trial schedule, we cannot eliminate the node and must explore the branch

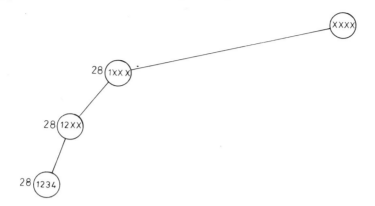

Fig. 7.2 The branch of the elimination tree explored first.

beyond it. If we arrive at a final node and find that the schedule there has C_{max} less than that of the trial, then this schedule becomes the new trial schedule. Eventually we will have eliminated or explored all the nodes and the trial schedule that remains must be optimal.

So let us return to our problem and apply this scheme. Fig. 7.2 shows the one branch of the elimination tree that we have explored so far. The numbers by the nodes are the lower bounds. Now the lower bound at 12XX is 28, which is no less than C_{max} of the trial. Thus we cannot improve upon the trial 1234 by exploring the branch to 1243. Similarly the lower bound at node 1XXX is also 28 and we may eliminate the branches to nodes 13XX and 14XX and beyond. Thus we have arrived at the stage shown in Fig. 7.3.

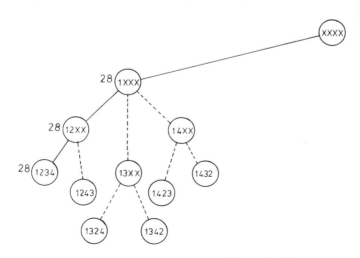

Fig. 7.3 All nodes beyond 1XXX fully explored.

Next we explore the branch XXXX to 2XXX and beyond. We make the following calculations.

At 2XXX $A = (J_2), U = \{J_1, J_3, J_4\}$.
                $\alpha_2 = 2, \beta_2 = 6, \gamma_2 = 11$.
                $lb(A) = \max\{2 + 10 + 10, 6 + 19 + 2, 11 + 14\}$
                      $= 27$.

This lower bound is less than 28, the $C_{\max}$ of the trial. Therefore we must explore the branches beyond the node 2XXX.

At 21XX $A = (J_2, J_1), U = \{J_3, J_4\}$.
                $\alpha_1 = 2 + 1 = 3, \beta_1 = \max\{2, 6\} + 8 = 14$,
                $\gamma_1 = \max\{14, 11\} + 4 = 18$.
                $lb(A) = \max\{2 + 9 + 10, 14 + 11 + 2, 18 + 10\}$
                      $= 28$.

At 23XX $A = (J_1, J_3), U = \{J_1, J_4\}$.
                $\alpha_3 = 2 + 6 = 8, \beta_3 = \max\{8, 6\} + 2 = 10$,
                $\gamma_3 = \max\{10, 11\} + 8 = 19$.
                $lb(A) = \max\{8 + 4 + 11, 10 + 17 + 2, 19 + 6\}$
                      $= 29$.

At 24XX $A = (J_2, J_4), U = \{J_1, J_3\}$.
                $\alpha_4 = 2 + 3 = 5, \beta_4 = \max\{5, 6\} + 9 = 15$,
                $\gamma_4 = \max\{15, 11\} + 2 = 17$.
                $lb(A) = \max\{5 + 7 + 10, 15 + 10 + 4, 17 + 12\}$
                      $= 29$.

The lower bounds at each of these nodes are not less than the value of $C_{\max}$ for the trial schedule; so there is no need to explore these branches further. Thus we arrive at the situation shown in Fig. 7.4.

Our next task is to explore the branches beyond 3XXX.

At 3XXX $A = (J_3), U = \{J_1, J_2, J_4\}$.
                $\alpha_3 = 6, \beta_3 = 8, \gamma_3 = 16$.
                $lb(A) = \max\{6 + 6 + 9, 8 + 21 + 2, 16 + 11\}$
                      $= 31$.

Since this lower bound is greater than 28, we explore no further along this branch. Finally we look at the branches beyond 4XXX.

At 4XXX $A = (J_4), U = \{J_1, J_2, J_3\}$.
                $\alpha_4 = 3, \beta_4 = 12, \gamma_4 = 14$.
                $lb(A) = \max\{3 + 9 + 9, 12 + 14 + 4, 14 + 17\}$
                      $= 31$.

Since this lower bound is greater than 28, there is no need to consider the branches beyond 4XXX.

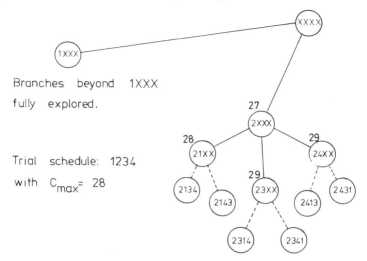

Branches beyond 1XXX
fully explored.

Trial schedule: 1234
with $C_{max}= 28$

Fig. 7.4 All nodes beyond 1**XXX** and 2**XXX** fully explored.

We have now explored the entire elimination tree and shown that no schedule can have a C_{max} less than 28. So the current trial schedule, viz. 1234 is optimal. The complete elimination tree for the above solution is given in Fig. 7.5.

In our solution here it happened that the first schedule we evaluate remained the trial for the entire solution. This happened purely by chance. Had we explored the branch **XXXX** to 4321 first, the trial schedule would have changed in the course of the solution. See Problem 7.6.1.

The type of search procedure we used above is called a **depth-first** search. In other words, in our search of the tree we selected a branch and systematically worked down it until we had either eliminated it on the grounds of a lower bound or had reached its final node, which had either become the trial schedule or was eliminated. This search strategy has the advantage that the computer need only remember the $\alpha_{i(k)}$, $\beta_{i(k)}$, $\gamma_{i(k)}$ and $lb(A)$ for the nodes in the branch currently being searched. Since there are at most $(n - 1)$ such nodes in an n job problem, this search procedure requires little storage. Table 7.1 below lists the nodes at which these quantities must be evaluated and remembered in order to eliminate certain other nodes in the tree.

Note that at no time was it necessary to remember the value of $\alpha_{i(k)}$, $\beta_{i(k)}$, $\gamma_{i(k)}$ and $lb(A)$ at more than $3 = 4 - 1$ nodes. Moreover, because of the systematic way in which we searched the tree, once the immediate need for these quantities at a node was past we knew we would never need them again.

We see then that this depth-first search requires very little storage. It

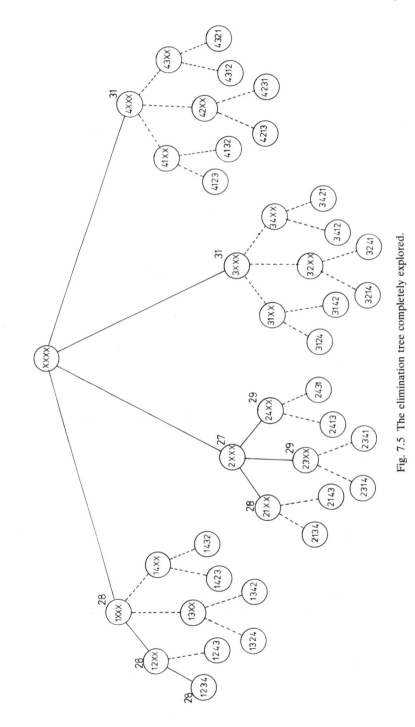

Fig. 7.5 The elimination tree completely explored.

Table 7.1—Memory requirements during the depth-first search

Nodes eliminated or fully searched	$\alpha_{i(k)}$, $\beta_{i(k)}$, $\gamma_{i(k)}$, and lower bounds required at nodes
1234	1XXX, 12XX, 1234
1243	1XXX, 12XX
1XXX and beyond	1XXX
21XX and beyond	2XXX, 21XX
23XX and beyond	2XXX, 23XX
24XX and beyond	2XXX, 24XX
3XXX and beyond	3XXX
4XXX and beyond	4XXX

may, however, require a great deal of computation. We were 'lucky'—well, I set the problem—in that the lower bounds eliminated many of the branches early on. It could have happened otherwise. We might have needed to explore the branches much further before they were eliminated. See Problem 7.6.1.

An alternative search strategy is a **frontier-search** or **branch-from-lowest-bound**. As the latter name suggests, in this we always branch from a node with the current lowest lower bound. It is perhaps easiest to follow this procedure through an example. Below we solve the $4/3/F/C_{max}$ problem again, but this time by a frontier-search. The calculations leading to the lower bounds either have been given already or are left to you as exercises.

First of all we simultaneously branch to each of the four nodes 1XXX, 2XXX, 3XXX, 4XXX and calculate the lower bounds there. Thus we obtain the result shown in Fig. 7.6. The node 2XXX has the lowest lower bound so we branch from this to nodes 21XX, 23XX, and 24XX to obtain the result shown in Fig. 7.7. Here we find two nodes 1XXX and 21XX

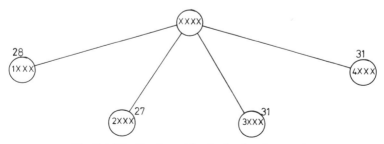

Fig. 7.6 The first branching in the frontier search.

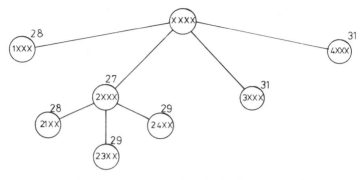

Fig. 7.7 The second branching in the frontier search.

share the same lower bound. Picking 21XX arbitrarily we branch to the nodes 2134 and 2143 as shown in Fig. 7.8. The schedule 2134 has $C_{max} = 28$. This is equal to the lowest of the lower bounds obtained in the tree. Thus 2134 is an optimal schedule.

(N.B. schedules 2134 and 1234 are both optimal with $C_{max} = 28$. Had we branched at 1XXX instead of 21XX the frontier-search would have found the same optimal schedule as the depth-first search.)

Had the node 2134 not turned out to be clearly an optimal schedule, we would have proceded as follows. Either schedule 2134 or 2143 would have been selected as the trial according to which has the smaller completion time. Then those nodes with lower bounds greater than this best yet C_{max} would have been eliminated from further searching. This being done, branching would have continued from a remaining node with the least lower bound. Continuing in the obvious way, an optimal schedule would eventually be found.

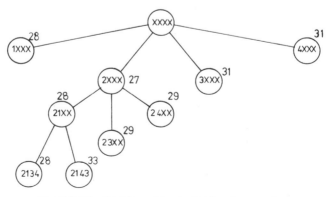

Fig. 7.8 The third branching in the frontier search.

In this example, the frontier search does not find an optimal schedule faster than the depth first search. However, see Problem 7.6.1. There you are asked to solve the above problem with a depth-first search that sweeps across the tree from right to left rather than from left to right. You will find in doing so that you need more calculation than in the above frontier search. In general, a frontier search will require less calculation than a depth-first search. The frontier search chooses which branch to explore next in a more intelligent fashion and so usually finds an optimal solution faster. Unfortunately this better performance is bought at a cost of greater storage requirements. Consider the memory requirements needed in the above.

Here we are required to keep track of $\alpha_{i(k)}$, $\beta_{i(k)}$, $\gamma_{i(k)}$ and the lower bound at as many as 7 nodes. The depth-first search only requires these quantities at most three nodes at a time. Table 7.2 should also make clear the reason for calling the branch-from-lowest-bound search a frontier search. At such stage it must remember the lower bounds of the non-eliminated nodes on the frontier of the nodes explored so far.

Before closing this section, notice that the above example does not conform to the conditions (5.2) under which Johnson's Algorithm may be applied. If we apply the algorithm nonetheless, we obtain the schedule 2314, which by the lower bound at 23XX must have $C_{\max} \geq 29$ and so cannot be optimal. It is clear, therefore, that conditions (5.2) are not redundant, but are necessary to the success of that algorithm.

Table 7.2—Memory requirements during the frontier search

Branching	'Frontier' of nodes at which $\alpha_{i(k)}$, $\beta_{i(k)}$, $\gamma_{i(k)}$ and the lower bound are required					
1st	1XXX	2XXX	3XXX	4XXX		
2nd	1XXX	21XX	23XX	24XX	3XXX	4XXX
3rd	1XXX	2134	2143	23XX	24XX	3XXX 4XXX

7.4 SOME GENERAL POINTS ABOUT THE BRANCH AND BOUND APPROACH

Branch and bound has been used to solve so many classes of scheduling problems that it deserves to be treated in some generality. This we now do. We begin by studying the logical structure underlying its search of the elimination tree. Indeed, first let us consider the construction of the elimination tree itself.

The first requirement is that we should be able to repeatedly partition the set Ω of all possible schedules into smaller and smaller subsets until at last we obtain a partition of Ω into subsets which each contain one and only one schedule. Thus, if Ω is first partitioned into $\Omega_1, \Omega_2, \ldots, \Omega_\nu$ and then each Ω_i into $\Omega_{i_1}, \Omega_{i_2}, \ldots$, and so on, we obtain the tree structure shown in Fig. 7.9.

As above we call each subset in this tree a node. In the example of the last section it may not have been obvious that nodes corresponded to subsets of schedules, but they did. For instance, the subset Ω_1 was labelled 1XXX and was the subset of schedules with job 1 first in the processing sequence. Similarly Ω_{23} was labelled 23XX and was the subset of schedules with job 2 first and job 3 second in the processing sequence. And so on.

Apart from this branching structure of subsets the method also requires that we have a bounding function lb. Thus, if Y is the subset corresponding to a particular node, we should be able to calculate a lower bound $lb(Y)$ on the performance measure for all the schedules in Y. If $c(y)$ is the value of the performance measure for a schedule y, then we require, therefore,

$$lb(Y) \leq c(y) \qquad \text{for all } y \text{ in } Y.$$

We need two further components to define a general branch and bound procedure. Firstly there is the trial schedule, y^*. This may not be set initially, but at some point in the procedure it becomes set to the schedule which was the best value of the performance measure found so far. Secondly we need a search strategy, which, not surprisingly, tells us how to search the tree.

Consider how a branch and bound procedure progresses towards a solution. At any stage the nodes are partitioned into three classes.

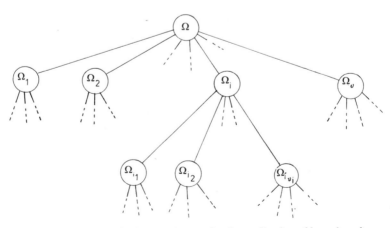

Fig. 7.9 The tree of subsets necessary for the application of branch and bound.

Z_1—the class of nodes which have either been eliminated or fully explored. A **fully explored** node is either a terminal node for which the performance measure has been evaluated exactly or a node in the main body of the tree such that lower bounds have been evaluated at all the nodes immediately beyond it.

Z_2—the class of **partially explored** nodes. The lower bound has been evaluated for any node in this class, but the node itself has been neither eliminated nor fully explored.

Z_3—the class of **unexplored nodes**. These nodes neither have been eliminated implicitly by the elimination of a node that preceeds them in the tree nor have been examined so far by the search procedure.

For instance, in the frontier search solution of the previous section at the stage displayed in Fig. 7.7 the sets Z_1, Z_2, and Z_3 are:

$$Z_1 = \{\text{XXXX, 2XXX}\};$$

$$Z_2 = \{\text{1XXX, 21XX, 23XX, 24XX, 3XXX, 4XXX}\};$$

and the remaining 33 nodes all lie in Z_3.

A branch is said to be **fathomed** when there is no further reason to search along it. This happens in one of two ways. Either a node and, hence, all the subsequent nodes are eliminated or the terminal node is reached and the performance measure evaluated there.

At the beginning of a branch and bound solution all the nodes lie in class Z_3. As the solution progresses nodes are moved either directly to class Z_1 or via the intermediate class Z_2. The procedure ends with all the nodes in class Z_1, all branches fathomed, and the knowledge that the current trial schedule is optimal.

We now examine a stage of a branch and bound solution to see exactly how nodes are moved from Z_3 into Z_1. The search procedure selects a pair of nodes Y and Y_i such that Y_i lies directly beyond Y in the tree and, moreover, Y lies in Z_2 and Y_i in Z_3. In other words, the tree branches from node Y to nodes $Y_1, Y_2, \ldots, Y_v$; Y_i is one of those subsequent nodes, which has yet to be explored. There is a slight difficulty when we consider the first stage. Initially all the nodes lie in Z_3; Z_2 is empty. Thus the search procedure will not be able to select a suitable node Y. To avoid this problem the first stage is a conventional one, in which the initial node Ω representing the set of all possible schedules is placed in Z_2.

When Y and Y_i have been selected, there are two possibilities: either Y_i contains more than one schedule or Y_i specifies a schedule uniquely, i.e. $Y_i = \{y\}$. We consider each of these possibilities in turn.

Y_i contains more than one schedule. We calculate $lb(Y_i)$. If $lb(Y_i) \geq c(y^*)$, the best value of the performance measure yet, then we eliminate node Y_i and all the nodes that lie beyond it. Thus they are

moved directly from Z_3 to Z_1. If $lb(Y_i) < c(y^*)$, we cannot eliminate Y_i and so it moves from Z_3 to Z_2.

$Y_i = \{y\}$, *i.e. Y_i contains exactly one schedule.* Here we calculate $c(y)$. If $c(y) \geq c(y^*)$, we eliminate Y_i and thus move it from Z_3 to Z_1. If $c(y) < c(y^*)$, or if the trial schedule has not been set, then y becomes the new trial schedule. In this case we must examine all the nodes in Z_2 to see if they are eliminated by the value of the performance measure for this new trial schedule. If such a node is eliminated, we move it together with all the nodes beyond it into Z_1.

In either of the above cases the examination of Y_i may complete the exploration of the node Y. If so we move node Y from Z_2 to Z_1.

A little thought shows that for any sensible search strategy the branch and bound procedure will eventually terminate with all the nodes in Z_1 and the trial schedule that remains will be optimal.

In the last section we met two search strategies: the depth-first and the frontier. The depth-first strategy selects as Y the node most recently placed in Z_2. (For this reason it is also called the **newest-active-node** search.) This strategy will usually select as Y the Y_i of the previous stage, thus progressing node by node down a branch. However, if the previous stage fathomed the branch, the search retreats up the branch until it encounters a node that has not been fully explored; it then searches one of the unfathomed branches beyond this node. The frontier search strategy is more complicated. When completely free to choose, it selects as Y a node in Z_2 which has the least lower bound. This choice fixes Y until it has been completely explored. Thus, if the tree branches from Y to the subsequent nodes Y_1, Y_2, ..., Y_ν, the next ν stages of the search are fixed as Y_1, Y_2, ..., Y_ν are examined.

We have seen that dynamic programming has two major faults. First, it requires that very many quantities be remembered during the solution. Second, although much faster than complete enumeration, it may nonetheless take prohibitively long to solve medium to large problems. Is branch and bound subject to the same disadvantages? We consider storage requirements first.

The maximum size of Z_2 determines the storage requirements of a branch and bound procedure. From its definition it is clear that at any stage the procedure must remember lower bounds and other intermediate quantities at and only at those nodes currently in this class. No quantities will have been calculated for any node in Z_3; whereas the lower bounds and other quantities at the nodes in Z_1, although once calculated, will have outlived their use and will have been forgotten. As we saw in the last section, in a depth-first search Z_2 will never be larger than the number of nodes in the largest branch of the elimination tree. This is much, much less than the storage requirements of an equivalent dynamic programming solution. For a frontier search the size of Z_2 is completely unpredictable. In theory, the size of Z_2 may become extremely large during the solution of some problems. In practice, however, this does not seem to happen and frontier searches may be used with only a small risk that they will exhaust the available storage capacity.

The number of operations required, and hence the time required, to solve a problem by branch and bound is unpredictable, whatever search strategy is used. It might happen that the procedure has to explore fully virtually every node, in which case it would take as long as complete enumeration. Indeed, it might take longer because branch and bound involves more computation per node than complete enumeration. Nonetheless, in general branch and bound does perform a great deal better than complete enumeration. But it should not be assumed from this that it can solve any problem in practice. Theoretically, like dynamic programming, it always finds an optimal solution, but it may take prohibitively long to do so. For instance, a $10/10/G/B$ problem posed by Muth and Thompson (1963) still has to be solved optimally (Graham *et al.*, 1979).

Obviously the search strategy used is an important determinant of the time required to solve a problem. As we have already noted, a frontier search generally finds an optimal schedule faster than a depth-first search. Equally important is the quality of the lower bounds. If the lower bound at a node is good, i.e. not much less than the least value of the performance measure there, then the procedure generally finds an optimal schedule after examining fewer nodes than it would with poorer bounds. Good bounds eliminate nodes high up in the tree, thus reducing the search substantially. Consider the example of the previous section. At each node zero is a lower bound, albeit a much poorer one than (7.5). When zero is used, every branch must be explored. Clearly the better bound (7.5) is to be preferred. Despite these comments, it must not be assumed that the better the lower bounds the faster will branch and bound find the optimal schedule. It depends upon how long the calculation takes on each node. it may be faster overall to calculate poor lower bounds quickly and examine many nodes in the search than to calculate good lower bounds slowly and examine far fewer nodes. However, empirical investigations have shown that generally the extra effort required to calculate good lower bounds is time well spent; the rate of eliminating nodes increases more than enough to compensate. For instance, Baker (1975) has compared the performance of a branch and bound algorithm for $n/m/P/F_{max}$ problems based upon bounds similar to (7.5) with one based upon a more complex bound of McMahon and Burton (1967). (See Problem 7.6.6.) He found that for simple problems the extra calculation for the second bound was not worthwhile, but as the problems became more difficult the second procedure become substantially better and was, on balance, the one to be recommended. Rinnooy Kan *et al.* (1975) have suggested using different bounds in different parts of the elimination tree. When a node high in the tree is eliminated, many subsequent nodes are eliminated at the same time. Lower down in the tree each node has fewer subsequent nodes to be eliminated along with it. Thus the best procedure might be to spend time calculating very good bounds in the upper levels of the tree, but calculate quick but poor bounds in the lower levels.

Apart from the choice of search strategy and lower bounds, there are other ways in which we may attempt to increase the speed of finding an optimal solution. Firstly, we may prime the procedure with a near-optimal schedule as the first trial. The better the first trial the more nodes we may expect to be eliminated in the early stages. Therefore, assuming that we can find a near optimal schedule, we may save ourselves much computation. That assumption, of course, begs an awkward question, but it is not an insurmountable one: how do we find a near optimal schedule? The answer is to use heuristic methods. We shall discuss these at length in Chapters 10 and 11; but, briefly, these are a family of sensible 'rules of thumb' which we expect from experience and intuition to produce good, if not optimal, schedules. For instance, it is known that, if Johnson's Algorithm is applied to an $n/3/F/F_{max}$ problem for which conditions (5.2) do not hold, then nonetheless a good schedule is obtained. Thus we might use this to generate a first trial schedule for a branch and bound solution. Both Baker (1975), and Kohler and Steiglitz (1976) report that this tactic of priming the search with a good trial schedule can be most successful.

Secondly, we may consider employing dominance conditions. We have already suggested their use in dynamic programming; their use in branch and bound is much better investigated. Suppose we have a set of conditions such that, when they apply to a pair of nodes, we may deduce that all the schedules at one node can do no better than the best schedule at the other. Then clearly we may eliminate the first node from further consideration. For example, consider an $n/3/F/F_{max}$ problem and refer to the notation used in developing the recursions (7.1). Suppose that we compare two nodes $J_{i(1)}, J_{i(2)}, \ldots, J_{i(K)}, XX \ldots X$ and $J_{j(1)}, J_{j(2)}, \ldots, J_{j(K)}, XX \ldots X$ at which the same K jobs have been assigned to the first K positions, but in different orders; i.e. the nodes share the same set of unassigned jobs. Suppose further that the following conditions hold:

$$\alpha_{i(K)} \geq \alpha_{j(K)}, \quad \beta_{i(K)} \geq \beta_{j(K)}, \quad \gamma_{i(K)} \geq \gamma_{j(K)};$$

i.e. the completion time of the Kth job on each machine is no earlier under the first subsequence than it is under the second. Then, however we complete the processing of the unassigned jobs at the first node, we shall do at least as well with the same completion at the second. Thus we may eliminate the first node from the search. Checking for such dominance conditions during the search of a tree may take considerable computation and storage. Nonetheless, since they may eliminate many nodes before lower bounding arguments can do so, their use may curtail the search sufficiently that, overall, a reduction in computational requirements is obtained. Indeed, this has been found in practice. Baker (1975), Rinnooy Kan *et al.* (1975), and Lageweg *et al.* (1978) all report that careful inclusion of dominance conditions in branch and bound can lead to improvements in performance.

A final method for saving computation is to accept a sub-optimal solution. Suppose we agree that any solution within 10% of the optimal would be satisfactory; i.e. if y_{opt} is an optimal schedule and y is a schedule such that $c(y) \leqslant 1.10\, c(y_{opt})$, then we would accept y as a solution. Now suppose that at some stage of a branch and bound search we have found a trial schedule y^*. Then we may eliminate all nodes with lower bounds greater than $c(y^*)/1.10$. For, if one of those nodes were to lead to an optimal schedule, then it would have to be within 10% of our current trial. Thus when looking for a sub-optimal solution we may eliminate nodes faster than if we were only prepared to accept an optimal schedule. Kohler and Steiglitz (1976) report that the savings in computation that result may be dramatic.

7.5 REFERENCES AND FURTHER READING

Conway, Maxwell, and Miller (1967), and Baker (1974) both contain introductions to the use of branch and bound in scheduling at a level comparable with our presentation. Lenstra (1977) and Rinnooy Kan (1976) discuss the subject in much greater technical detail.

As remarked earlier, branch and bound methods have been the most successful of the non-heuristic approaches for solving scheduling problems—although there are occasions when a dynamic programming solution is to be preferred, e.g. see Baker and Schrage (1978). Given this success, it may seem strange that we have only studied a branch and bound solution of one problem, namely the $n/3/F/F_{max}$ one. The reason is simple. We noted that the quality of the lower bounds in as important determinant of performance of a procedure. Hence, much skill has gone into the design of these. In consequence, the resulting bounds are based upon mathematical theory far beyond the level of that assumed in this book. The following references discuss three important applications of branch and bound to scheduling, but they do require much mathematical knowledge and maturity on the part of the reader. Rinnooy Kan *et al.* (1975) consider the $n/1//\Sigma \gamma_i$ problem; Lageweg *et al.* (1977) consider the $n/m/G/C_{max}$ problem; and Lageweg *et al.* (1978) is the most up to date reference on the permutation flow-shop problem $n/m/P/C_{max}$, their theory superceding that discussed in Section 7.3 (see also Bansal, 1979; and Potts, 1978). Burkard (1980) discusses a general approach to developing lower bounds for job shop problems.

In Section 7.4 we discussed the general structure of branch and bound, and also how we might improve the speed with which a solution is found. Kohler and Steiglitz give a much more detailed and advanced treatment of these ideas, supporting their discussion with both empirical and theoretical results. The theoretical performance of branch and bound is also discussed by Lenstra and Rinnooy Kan (1978).

7.6 PROBLEMS

1. Solve the example of Section 7.3 by a depth-first search that sweeps from right to left, i.e. start with the branch from XXXX to 4321.
2. Solve the following $4/3/F/C_{max}$ problem with zero ready times
   (i) by a depth-first search,
   (ii) by a frontier search.

Job	Processing Times		
J_i	a_i	b_i	c_i
1	6	9	3
2	14	8	2
3	3	17	5
4	10	10	10

3. Explain why the method of Section 7.3 cannot be extended to solve $n/4/F/C_{max}$ problems. Extend the method to solve the $4/4/P/C_{max}$ problem with zero ready times and processing times:

Job	Processing Times			
J_i	a_i	b_i	c_i	d_i
1	6	3	8	2
2	9	4	3	7
3	3	8	4	6
4	3	4	8	6

N.B. here d_i are the processing times on the fourth machine, not due dates.
   4. Extend the method of Section 7.3 to allow for non-zero ready times.
Solve the $n/3/P/C_{max}$ problem with data:

Job J_i	Ready time r_i	Processing Times a_i	b_i	c_i
1	0	13	5	4
2	6	7	3	16
3	14	16	9	18
4	12	4	8	2

Could you extend your method to solve an $n/3/F/C_{max}$ problem with non-zero ready times?
   5. In Section 7.3 we used the lower bound:

$$lb(A) = \max\{q_1, q_2, q_3\},$$

where

$$q_1 = \alpha_{i(K)} + \sum_{J_i \text{ in } U} a_i + \min_{J_i \text{ in } U}\{b_i + c_i\},$$

$$q_2 = \beta_{i(K)} + \sum_{J_i \text{ in } U} b_i + \min_{J_i \text{ in } U}\{c_i\},$$

$$q_3 = \gamma_{i(K)} + \sum_{J_i \text{ in } U} c_i.$$

Show that this bound may be improved to

$$lb'(A) = \max\{q_1, q_2', q_3'\}$$

where

$$q_2' = \max\{\beta_{i(K)}, \alpha_{i(K)} + \min_{J_i \text{ in } U}\{a_i\}\} + \sum_{J_i \text{ in } U} b_i + \min_{J_i \text{ in } U}\{c_i\}$$

$$q_3' = \max\{\gamma_{i(K)}, \beta_{i(K)} + \min_{J_i \text{ in } U}\{b_i\}, \alpha_{i(K)} + \min_{J_i \text{ in } U}\{a_i + b_i\}\}$$
$$+ \sum_{J_i \text{ in } U} c_i$$

Use this form of the lower bound to solve the $4/3/F/C_{max}$ problem with data:

Job J_i	Processing Times		
	a_i	b_i	c_i
1	25	10	4
2	21	18	6
3	20	12	9
4	16	19	3

6. Show that the following is also a lower bound at the node $J_{i(1)}, J_{i(2)}, \ldots, J_{i(K)}, XX \ldots X$ for the $n/3/F/C_{max}$ problem.

$$lb''(A) = \alpha_{i(K)} + \max_{J_i \text{ in } U}\{a_i + b_i + c_i + \sum_{\substack{J_j \text{ in } U \\ J_j \neq J_i}} \min\{a_j, c_j\}\}$$

7. Consider the $n/1//C_{max}$ problem with sequence dependent processing times as discussed in Problem 6.6.7. Develop a branch and bound solution as follows. Notice that the elimination tree will have the same structure as that discussed in Section 7.2. At node $J_{i(1)}, J_{i(2)}, \ldots, J_{i(K)}, XX \ldots X$ show a lower bound is

$$lb(A) = \pi_{0i(1)} + \pi_{i(1)i(2)} + \ldots + \pi_{i(K-1)i(K)} + (n - K)\pi_{min}$$

where

$$\pi_{min} = \min_{\text{all } i, j}\{\pi_{ij}\}.$$

Hence solve the $4/1//C_{max}$ problem set in Problem 6.6.7.

8. Find a suboptimal solution within 15% of the optimal for the $6/3/F/C_{max}$ problem with data:

Job J_i	Processing Times		
	a_i	b_i	c_i
1	6	10	11
2	8	12	12
3	10	14	10
4	10	10	9
5	6	12	8
6	8	14	10

You may find it useful to set your first trial solution to that which Johnson's Algorithm would determine if it was applicable.

9. Solve the $n/1//\overline{T}$ problem discussed in Section 6.2 by branch and bound.

Chapter **8**

Integer Programming Formulations

8.1 INTRODUCTION

There is a sizeable body of literature which suggests solving scheduling problems by recasting them as mathematical programmes, particularly integer programmes. (A mathematical programme is simply a general form of constrained optimisation problem. See below.) These recast problems may be solved by standard algorithms, which have been developed for solving general mathematical programmes. Hence, translating back, we obtain optimal schedules. Stated simply, this sounds a very promising approach; in practice it is not. The standard mathematical programming algorithms are practically applicable only to small problems; the recast scheduling problems can be very large indeed. Thus the *recasting* is literally just that; the inherent difficulties are rephrased, but not into a more tractable form. Empirically this confirms that scheduling problems are in general very difficult—and do not just appear to be so.

In the next section we translate an $n/m/P/C_{max}$ scheduling problem into a mixed integer programme. We shall not discuss methods of solving such problems, but instead refer to the literature. In the sense that I have indicated, such reformulations are misleading since the integer programmes are no easier to solve than the original problems. Nonetheless, no introductory treatment of scheduling would be complete without some mention of the following. Moreover, the modelling techniques used are of considerable importance elsewhere in operational research.

The body of theory and algorithms called *mathematical programming* deals with problems of the following kind:

$$\text{Minimise} \quad f(x_1, x_2 \ldots x_l)$$

with respect to $x_1, x_2, \ldots x_l$ subject to the constraints:

$$g_1(x_1, x_2, \ldots, x_l) \leq b_1$$
$$g_2(x_1, x_2, \ldots, x_l) \leq b_2$$
$$\vdots$$
$$g_k(x_1, x_2, \ldots, x_l) \leq b_k$$

In other words, mathematical programming is a family of techniques for optimising a function subject to constraints upon the independent variables. In scheduling we wish to optimise a performance measure subject to technological constraints on the allowable processing order. Thus it is not surprising that, given a little ingenuity, the latter can be translated into the former type of problem; both concern optimisation under constraints.

In **integer programming** or, to be strictly correct, **mixed integer programming** some of the independent variables are constrained to be integral. Often they are only allowed the values 0 or 1 and are used to indicate the absence or presence of some property. Furthermore, the functions $f, g_1, g_2, \ldots g_k$ are linear. Thus the standard problem takes the form:

$$\text{minimise} \quad c_1 x_1 + x_2 x_2 + c_3 x_3 + \ldots + c_l x_l$$

subject to
$$g_{11} x_1 + g_{12} x_2 + g_{13} x_3 + \ldots + g_{1l} x_l \leq b_1$$
$$g_{21} x_1 + g_{22} x_2 + g_{23} x_3 + \ldots + g_{2l} x_l \leq b_2$$
$$\vdots$$
$$g_{k1} x_1 + g_{k2} x_2 + g_{k3} x_3 + \ldots + g_{kl} x_l \leq b_k$$

and certain of the $x_1, x_2, \ldots x_l$ are limited to integral values. Note also that some of the inequality constraints may be replaced by strict equality. Methods of solving such problems are reviewed in Garfinkel and Nemhauser (1972) and most introductory texts on operational research e.g. Daellenbach and George (1979). By and large, those methods can be classified as either implicit enumeration, in particular branch and bound, or cutting plane. Both require much computation. Moreover, both are based upon properties of integer programmes in general and pay no regard to particular properties of the problem being solved. As a result they tend to take longer to find a solution than implicit enumeration algorithms designed specifically for a particular class of problems. For example, consider two methods of solving a scheduling problem. Firstly, we may translate it into an integer programme and solve that by branch and bound with the bounds based upon general integer programming theory. Or, secondly, we may tackle the problem directly by branch and bound with bounds based upon our knowledge if the physical properties of schedules. The lower bounds found in the first case are usually poorer than those found in the second case and the branch and bound search correspondingly longer. So, as we have said, it is better to approach scheduling problems directly rather than indirectly via integer programming.

Assumptions for Chapter 8
There are no specific restrictions upon the application of integer programming methods.

8.2 WAGNER'S INTEGER PROGRAMMING FORM OF $n/m/P/C_{max}$ PROBLEMS

Wagner (1959) has introduced the following integer programming formulation of the permutation flow-shop problem $n/m/P/C_{max}$. We assume that all ready times are zero.

As usual the subscript i will refer to the job J_i; the subscript j to the machine M_j; and the subscript k to the kth position in the processing sequence. Thus the jobs will be processed through each machine in the order $(J_{i(1)}, J_{i(2)}, \ldots, J_{i(k)}, \ldots, J_{i(n)})$. Moreover, since the permutation flow-shop is a specialisation of the flow-shop, each job has technological constraints $(M_1, M_2, \ldots, M_m)$.

To model this problem we introduce n^2 variables constrained to take the values 0 or 1:

$$X_{ik} = \begin{cases} 1, & \text{if } J_i \text{ is scheduled in the } k\text{th position of the processing sequence;} \\ 0, & \text{otherwise.} \end{cases}$$

The constraints that must be obeyed by these variables are:

$$\sum_{i=1}^{n} X_{ik} = 1 \quad \text{for} \quad k = 1, 2, \ldots, n, \tag{8.1}$$

i.e. exactly one job is scheduled in the kth position; and

$$\sum_{k=1}^{n} X_{ik} = 1 \quad \text{for} \quad i = 1, 2, \ldots, n, \tag{8.2}$$

i.e. each job is scheduled in exactly one position. These constraints force each X_{ik} to take the values 0 and 1 only, provided we also demand that the X_{ik} are non-negative integers.

Next we introduce two sets of non-negative real variables; I_{jk} and W_{jk}.

The first set, I_{jk}, represent the idle times on the machines; the second set, W_{jk}, the waiting times for the jobs between machines. The use of subscripts here will be slightly different to that introduced in Section 1.5. Specifically I_{jk} is the idle time of machine M_j between the completion of $J_{i(k)}$ in the processing sequence and the start of $J_{i(k+1)}$. Thus I_{jk} is defined for $j = 1, 2, \ldots, m$ and $k = 1, 2, \ldots, n - 1$. Moreover, since there need be no idle time on the first machine, $I_{1k} = 0$ for $k = 1, 2, \ldots, n - 1$. The waiting time W_{jk} is the time that $J_{i(k)}$ must spend between completion on M_j and starting to be processed on M_{j+1}. Thus W_{jk} is defined for $j = 1, 2, \ldots, m - 1$ and $k = 1, 2, \ldots, n$. Moreover, since there is nothing to delay the first job processed, $W_{j1} = 0$ for $j = 1, 2, \ldots, m - 1$. These quantities are indicated in Fig. 8.1. Note that $I_{j+1,k}$ and $W_{j,k+1}$ cannot both be non-zero for a semi-active schedule; they are shown as such in the diagram purely for illustrative purposes.

The next set of constraints that we introduce simply say that the times t_{jk} between completion of the job $J_{i(k)}$ on M_j and the start of the job $J_{i(k+1)}$ on M_{j+1} must be well defined. From Fig. 8.1 it is clear that

$$t_{jk} = T_{jk} + p_{i(k+1),j} + W_{j,k+1}$$
$$= W_{jk} + p_{i(k),j+1} + I_{j+1,k} \qquad (8.3)$$

We must express $p_{i(k+1),j}$ and $p_{i(k),j+1}$ in terms of the $X_{i(k)}$, because we do not know explicitly how to find the subscripts $i(k + 1)$ and $i(k)$. Now, since $X_{i,k+1}$ is zero except for $i = i(k + 1)$ when it is 1, it follows that

$$p_{i(k+1),j} = \sum_{i=1}^{n} X_{i,k+1} p_{ij} \qquad (8.4)$$

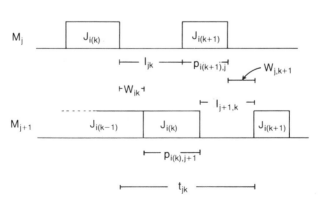

Fig. 8.1 The relationships between the variables in the integer programme.

Similarly

$$p_{i(k),j+1} = \sum_{i=1}^{n} X_{ik} p_{i,j+1} \tag{8.5}$$

Thus we may rewrite (8.3) as:

$$I_{jk} + \sum_{i=1}^{n} X_{i,k+1} p_{ij} + W_{j,k+1} - W_{jk} - \sum_{i=1}^{n} X_{ik} p_{i,j+1} - I_{j+1,k} = 0 \tag{8.6}$$

Clearly these constraints hold for $j = 1, 2, \ldots, m - 1$, $k = 1, 2, \ldots,$ $(n - 1)$. The constraints (8.6) not only ensure that the times t_{jk} are well defined, but also ensure that the technological constraints of the flow-shop are obeyed.

The constraints (8.1), (8.2), and (8.6) together with the demands that X_{ik} are non-negative integers and that W_{jk} and I_{jk} are non-negative real r.umbers, form the entire constraint set. The objective function, which we wish to minimise, is defined as follows.

Minimising C_{max} is equivalent to minimising the idle time on the last machine. The total idle time on M_m is given by the sum of the inter-job idle times I_{mk} plus the idle time that must occur before job $J_{i(1)}$ can start processing on M_m. Thus we seek to minimise

$$\sum_{k=1}^{n-1} I_{mk} + \sum_{j=1}^{m-1} p_{i(1),j}.$$

However, we must express the sum $\sum_{j=1}^{m-1} p_{i(1),j}$ in terms of X_{ik} variables. So we seek to minimise (c.f. (8.4) and (8.5))

$$\sum_{k=1}^{n-1} I_{mk} + \sum_{j=1}^{m-1} \left(\sum_{i}^{n} X_{i1} p_{ij} \right) \tag{8.7}$$

Thus our problem is to minimise (8.7) subject to constraints (8.1). (8.2) and (8.6) plus the non-negativity of the X_{ik}, W_{jk} and I_{jk}, and also the condition that the X_{ik} must be integral.

The speed with which integer programmes can be solved depends upon the number of variables and constraints in the problem. So it is informative to count these in Wagner's formulation. Counting the variables first, there are

n^2	integer variables	X_{ik},
$(m - 1)(n - 1)$	real variables	I_{jk} (N.B. $I_{1k} = 0$ for all k),
and $(m - 1)(n - 1)$	real variables	W_{jk} (N.B. $W_{j1} = 0$ for all j),

$\Rightarrow n^2 + 2(m - 1)(n-1)$ variables in total.

Next counting the constraints, there are

n	constraints of type (8.1),
n	constraints of type (8.2),
and $(m - 1)(n - 1)$	constraints of type (8.6),
$\Rightarrow$ $\overline{mn + n - m + 1}$	constraints in total.

There are, of course, $n^2 + 2(m - 1)(n - 1)$ non-negativity constraints upon the variables, but these need not be counted explicitly since integer programming algorithms invariably include them implicitly.

Practical applications of the above ideas have been disappointing, thus supporting the claim above that there is no advantage in translating scheduling problems into integer programming problems. Giglio and Wagner (1964) solved six $6/3/P/C_{max}$ problems using an integer programming algorithm due to Gomory. The number of computations required was either roughly equal to or greater than that required by complete enumeration for the original scheduling problems. See also Story and Wagner (1963).

8.3 REFERENCES AND FURTHER READING

Both Conway, Maxwell and Miller (1967) and Baker (1974) discuss integer programming formulations of scheduling problems. Rinnooy Kan (1976) gives the most recent survey, and also mentions some work of Nepomiastchy in which non-linear mathematical programming is used to obtain an approximate solution to general job-shop problems. Apart from Wagner (1959) other simple integer programming formulations of scheduling problems are given by Bowman (1959), Dantzig (1960), and Manne (1960). Greenberg (1968) presents a branch and bound solution to an integer programming version of $n/m/G/C_{max}$ problems. Although his approach is naive by present day standards, his results indicate the poorness of branch and bound solutions to integer programmes compared with a similar solution of the original scheduling problems.

Although I have been emphatic in my dismissal of integer programming approaches, let me grant them a use, at least in the immediate future. In Chapter 4 we discussed efficiency ideas. Recently Huckert *et al.* (1980) have used multiple-objective integer programming to find efficient schedules. In the short term recourse to such methods may be very productive. However, in the long term I am sure that purpose-built algorithms will be developed; the economics of computation will ensure that.

8.4 PROBLEMS

1. Reformulate the following $3/3/P/C_{max}$ problem as an integer programme:

	Processing Time		
Job	M_1	M_2	M_3
1	6	8	3
2	9	5	2
3	4	8	17

2. Show that the $n/m/P/\bar{F}$ problem may be reformulated as an integer programme by using the constraints above together with the objective function

$$\sum_{K=1}^{n}\left(\sum_{j=1}^{m-1} W_{jK} + \sum_{k=1}^{K} \sum_{i=1}^{n} X_{ik}p_{i1}\right)$$

Chapter **9**

Hard Problems and NP-Completeness

9.1 INTRODUCTION

By now it should be apparent that when there is a choice constructive algorithms are preferable to implicit enumeration. However, we have found constructive techniques for only the simplest of problems. For the rest we must use implicit enumeration—unless, of course, we find ourselves unable to develop suitable dynamic programming recursions or lower bounds; in that case complete enumeration is unavoidable.

The examples in Chapters 6 and 7 show that such solutions can involve much computation. Indeed, we estimated that at a rate of 1 mathematical operation per micro-second a dynamic programming solution of a $40/1/\bar{T}$ problem would need more than 4 years. We did not make a similar estimate for branch and bound, because its performance on particular problems is so unpredictable. Nonetheless, we can quote an empirical observation on its average performance. Ignall and Schrage (1965) report that their algorithm for the $n/3/F/C_{max}$ problem requires on average about twice as much time for $(n + 1)$ jobs as it does for n. Thus, if it takes 1 sec. to solve an n job problem, it will take 2^r sec. to solve an $(n + r)$ one. For $r = 20$, 2^r sec. = 12 days.

These formidable computational requirements provide the greatest possible encouragement to use constructive methods whenever possible. Moreover, we should surely devote our efforts to finding constructive algorithms for a wider range of problems. Sadly, it is probable that such efforts would be wasted. In this chapter I shall introduce you to a pessimistic, but mathematically fascinating conjecture—one which predicts that no constructive algorithms will ever be developed for the majority of scheduling problems. So, having forewarned you of the disappointment to follow, let us begin.

Assumptions for Chapter 9
We make no assumptions to limit the range of problems discussed, but we do henceforth assume that all data (i.e. processing times, due dates, etc.)

are integers. This, in fact, does not restrict us at all in practice because no data is ever known to more than a finite number of decimal places. Thus multiplying all the data in a problem by 10^l for some l will result in an essentially equivalent problem, but one in which all data are integers.

9.2 WHAT IS A 'GOOD' WAY OF SOLVING A PROBLEM?

Why do we prefer a constructive solution to an enumerative one? It is not really sufficient just to say because it requires less computation. Everyone would prefer to solve a $4/3/F/C_{max}$ problem by branch and bound than a $1000/1//\bar{C}$ problem by the SPT rule. Clearly we should say that we prefer constructive methods because for a *given size of problem* they need less computation. Our first objective is to make this statement more precise. Nonetheless, we shall continue to adopt a very informal approach. See Garey and Johnson (1979) for the formal underpinning of what is to follow.

We begin by being a little more careful about our use of the word *problem*. To date we have used it in two distinct ways: first, to describe general classes, e.g. the $n/m/G/\bar{T}$ problem; second, to refer to a particular case with particular data, e.g. the problem faced by Algy and friends (Section 1.1). From now on we shall only use **problem** in the sense of a class of problems. Particular cases will be referred to as **instances**. Thus in the following we shall not be concerned with the computation necessary to solve a given size of problem, but rather a given size of instance.

So what do we mean by the size of an instance? Computer scientists use the following definition and so shall we. Consider the description of $6/1//\bar{F}$ instance with processing times:

$$p_1 = 10, p_2 = 6, p_3 = 8, p_4 = 16, p_5 = 14, p_6 = 7.$$

Given that we know the problem type is $n/1//\bar{F}$, we need only specify the numbers

$$10, 6, 8, 16, 14, 7 \qquad (9.1)$$

to describe the problem precisely. We can find that $n = 6$ by counting the number in the list, and we can find that, say, $p_3 = 8$ by using the natural convention that the ith number is p_i. Note that the commas are essential. The string of digits 106816147, which results when they are omitted, is meaningless. Of course, we could replace the commas with colons, vertical strokes, or small pictures of elephants; but some **separator** symbol is necessary. In the string (9.1) there are 14 digits and separators. Because of this we say the instance has size 14.

Thus to determine the size of an instance we need to know:

(1) the type of problem that it is, that is its $n/m/A/B$ classification;
(2) the encoding convention used to list the data of the instance.

It would seem, therefore, that our definition of the size of an instance depends upon the encoding convention adopted. Indeed it does, for, if we list the processing times in binary form, we obtain

$$1010, 110, 1000, 10000, 1110, 111$$

and an instance size of 28. Equally we could have decided to include the redundant information that $n = 6$ in the list to save ourselves the trouble of counting the number of processing times. Thus we would have obtained

$$6, 10, 6, 8, 16, 14, 7$$

and an instance size of 16.

However, we shall see that this dependence upon convention is irrelevant to our needs, provided that we always adopt a reasonable encoding convention. By *reasonable* we mean the following.

(a) We do not inflate our list with unnecessary information or symbols. For example, we do not use repeated separators where one will do as in

$$10, , , 6, , , , 8, , , 16, 14, , , , 7.$$

Equally, we do not include such information as the date of Aunt Agatha's wedding in our list (unless, of course, that data is related to our problem).

(b) All numbers are given to some base other than 1. Thus a number K may be represented in binary or octal, or decimal; but not simply as a string of K 1's.

Suppose we consider two reasonable conventions for encoding instances of the same problem. It happens that there always exist two polynomials[†] $p(v)$ and $q(\eta)$ such that, if an instance has size v under one convention and η under another, then

$$\eta \leqslant p(v)$$
$$v \leqslant q(\eta) \tag{9.2}$$

For our purposes, this will be enough to make the size of instances effectively independent of the encoding convention that we use. (See Problems 9.6.1 and 9.6.7.)

The size of an instance is determined both by the number of data needed to define the instance and also by their magnitude. There is a dependence on their number because each quantity requires at least one digit to represent it. Hence the instance size must be greater than the number of data defining it. There is a dependence on the magnitudes of the data because the larger any one of them is the more digits will be required

---

[†]For reasons that will become apparent, all polynomials in this chapter are taken to be strictly increasing on the positive interval.

to represent it. But there is a difference between these dependencies. The first is linear, the second is logarithmic. Suppose that we have ten numbers all of roughly the same magnitude. Then to represent all of them we need about ten times as many digits as for any one. If we were to consider a hundred such numbers, we would need a hundred times as many digits. However, suppose that we consider two numbers, the first ten times as large as the second. Then we need only one more digit to represent the first than we do for the second. (At least, this is so if we represent numbers to the base 10; a similar observation can be made for other bases.) If the first number is a hundred times as big, we need only two more digits. Thus, whilst the size of an instance does depend both on the number of data and on their magnitude, the dependencies are not equal. We shall return to this point in the next section and again in Chapter 11.

Now we are in a position to discuss the relative computational merits of different algorithms. We shall do this by means of time complexity functions. The **time complexity function** $f(v)$ of an algorithm gives the maximum number of operations that would be required to solve an instance of size v.

In practice, it is unnecessary for us to determine time complexity functions completely; all we need is some indication of their behaviour as the problem size increases. Because of this, we introduce a very useful notation. We shall say that **f(v) is O(g(v))**—read 'f and g are of the same order'—whenever

$$f(v)/g(v) \to c, \text{ some constant, as } v \to \infty. \qquad (9.3)$$

Thus $f(v)$ is $O(g(v))$ if their ratio tends to become constant as v increases. In other words, as v increases the behaviour of $f(v)$ and $g(v)$ become more and more similar until they are essentially the same. As you will discover from Problems 9.6.4–5, the following properties hold.

If $f(v)$ is $O(v^n)$ and $g(v)$ is $O(v^m)$, then:

(i) $f(v) f(v)$ is $O(v^{n+m})$; $\qquad\qquad\qquad\qquad\qquad (9.4)$

(ii) $f(v)/g(v)$ is $O(v^{n-m})$; $\qquad\qquad\qquad\qquad\qquad (9.5)$

(iii) $f(g(v))$ is $O(v^{mn})$. $\qquad\qquad\qquad\qquad\qquad (9.6)$

Moreover the polynomial $(a_n v^n + a_{n-1} v^{n-1} + \ldots + a_o)$ is $O(v^n)$.

We shall say that an algorithm has **polynomial time complexity** if its time complexity function $f(v)$ is $O(p(v))$ for some polynomial $p(v)$. Otherwise an algorithm has **exponential time complexity**. Note that we include in our definition of exponential time complexity behaviour that is not normally regarded as exponential. For instance, if $f(v)$ is $O(v!)$ we say that it exhibits exponential behaviour.

The reason for making these definitions becomes apparent when we examine Tables 9.1 and 9.2 below. Table 9.1 compares the actual time

Table 9.1—The time requirements of algorithms with certain time complexity functions under the assumption that one mathematical operation takes one micro-second (Modified with permission from Garey and Johnson *Computers and Intractability: A Guide to the Theory of NP-Completeness.* W. H. Freeman and Company. Copyright © 1979. Figure 1.2, p. 7)

Time Complexity Function	v					
	10	20	30	40	50	60
v	0.00001 sec	0.00002 sec	0.00003 sec	0.00004 sec	0.00005 sec	0.00006 sec
v^2	0.0001 sec	0.0004 sec	0.0009 sec	0.0016 sec	0.0025 sec	0.0036 sec
v^5	0.1 sec	3.2 sec	24.3 sec	1.7 min	5.2 min	13 min
v^{10}	2.7 hrs.	118.5 days	18.7 yrs.	3.3 centuries	30.9 centuries	192 centuries
2^v	0.001 sec	1.0 sec	17.9 min	12.7 days	35.7 yrs	366 centuries
3^v	0.59 sec	58 min	6.5 yrs	3855 centuries	2×10^8 centuries	1.3×10^{13} centuries
$v!$	3.6 sec	770 centuries	8.4×10^{16} centuries	2.5×10^{32} centuries	9.6×10^{48} centuries	2.6×19^{66} centuries

Table 9.2—Increase in instance size solvable in a given time for a thousand-fold increase in computing speed. (Modified with permission from Garey and Johnson *Computers and Intractability: A Guide to the Theory of NP-Completeness.* W. H. Freeman and Company. Copyright © 1979. Figure 1.3, p. 8)

Time Complexity Function	Size of instance solved in a given time on slow computer	Size of instance solved in the same time on a computer 1000 times faster
v	v_1	$1000v_1$
v^2	v_2	$31.62v_2$
v^5	v_3	$3.98v_3$
v^{10}	v_4	$1.99v_4$
2^v	v_5	$v_5 + 10$
3^v	v_6	$v_6 + 6$
$v!$	v_7	$\begin{cases} v_7 + 3 & v_7 \leq 10 \\ v_7 + 2 & 10 < v_7 \leq 30 \\ v_7 + 1 & 30 < v_7 \leq 1000 \end{cases}$

requirements of several polynomial and exponential time complexity functions.

The first four rows of this table illustrate the time requirements for algorithms exhibiting polynomial behaviour; the last three illustrate exponential behaviour. It should be apparent that the increase in time requirements with instance size is far less dramatic for the polynomially bounded algorithms than for the others. Admittedly the case of v^{10} is somewhat intermediate, but nonetheless its requirements do not exhibit exponential growth. For instance, the time requirements of an $O(2^v)$ increase by a factor of $2^{50}(=10^{15})$ as v increases from 10 to 60; whereas those of an $O(v^{10})$ increase only by a factor of 6^{10} ($=6 \times 10^7$). In any case, it has been observed that in practice algorithms with polynomial time complexity tend to require $O(v^2)$ or $O(v^3)$; certainly $O(v^{10})$ is extremely rare, if indeed it has ever been observed. Thus empirically the distinction between polynomially and exponentially bounded algorithms is very marked.

It is also informative to consider how much larger an instance can be solved in a given time for a given increase in computing speed. To make this question precise, suppose we compare the performance of an algorithm on two computers. The slower one takes 1 micro-second to perform a mathematical operation; the faster one takes a nano-second (1 nanosecond = 10^3 microsecond). If in a given time we can solve an instance of size v on the first computer, how much larger an instance can we solve in the same time on the second? The answer is given in Table 9.2.

It can be seen that algorithms with polynomial time complexity allow a multiplicative increase in instance size for a given gain in computer power; whereas those with exponential time complexity only allow an additive increase in instance size. To emphasise this still further, in one minute the slower computer can solve an instance of size 36 with the v^5 time complexity algorithm and one of size 26 with the 2^v algorithm. In the same time the faster computer can solve an instance of size 143 with the v^5 algorithm and one of size 36 with the 2^v algorithm. Thus, if we have algorithms with polynomial time complexity, we know that advances in computer technology will enable us to solve much larger instances. However, if we only have solution techniques that are of exponential time complexity, we know that the gains from increased computing power will be slight.

For these reasons it has become conventional to talk of problems as being **well solved**, when we have developed an algorithm with polynomial time complexity. We are further encouraged in this definition because the existence of a polynomial time algorithm usually means that we have solved the problem through some insight, some understanding of the structure of solution. Exponential time algorithms are mainly types of exhaustive search procedures. The logic of the search may be very subtle and clever, but it seldom shows that we really understand the problem that we are trying to solve. You may appreciate this point from consideration of our earlier work. The constructive algorithms of Chapters 3, 4 and 5 are of polynomial time complexity, whereas dynamic programming and branch and bound algorithms have exponential time complexity. Perhaps we had better examine these assertions more closely.

As an example, we confirm that there is a polynomial time algorithm for constructing an SPT sequence in an $n/1//\overline{F}$ problem. Our aim is simply to sort the jobs into order of non-decreasing processing times. A straightforward way of doing this is as follows. We begin by listing the jobs in any order; for convenience we take this to be $(J_1, J_2, \ldots, J_n)$. We pass through the sequence comparing adjacent jobs, beginning with the jobs in the first and second positions. If their processing times are in decreasing order, i.e. if the former has the longer processing time, we interchange them. Otherwise, we leave them in their current positions. Next we compare the jobs in the second and third positions. (N.B. the job in the second position may have been placed there as a result of the first comparison and interchange.) If the processing times are in decreasing order, we interchange the second and third jobs. Otherwise we leave them. We continue through the sequence, comparing adjacent jobs. Thus we shall make $(n - 1)$ comparisons and at most $(n - 1)$ interchanges with the result that the longest job is moved down the sequence to the last position. For example, consider a $5\|1\|\|\overline{F}$ instance with data:

$$p_1 = 6, p_2 = 3, p_3 = 7, p_4 = 1, \quad \text{and} \quad p_5 = 3.$$

Applying this method we would make the interchanges shown in the first four rows of Table 9.3. Note that the longest job, viz J_3, is in the last

Table 9.3—Sorting the jobs into SPT order for the particular instance given in the text

Current Sequence	Compare Positions	Results
$(1,2,3,4,5)$	1st and 2nd	$p_1 = 6 > 3 = p_2$ so interchange
$(2,1,3,4,5)$	2nd and 3rd	$p_1 = 6 < 7 = p_3$ so leave
$(2,1,3,4,5)$	3rd and 4th	$p_3 = 7 > 1 = p_4$ so interchange
$(2,1,4,3,5)$	4th and 5th	$p_3 = 7 > 3 = p_5$ so interchange
$(2,1,4,5,3)$	1st and 2nd	$p_2 = 3 < 6 = p_1$ so leave
$(2,1,4,5,3)$	2nd and 3rd	$p_1 = 6 > 1 = p_4$ so interchange
$(2,4,1,5,3)$	3rd and 4th	$p_1 = 6 > 3 = p_5$ so interchange
$(2,4,5,1,3)$	1st and 2nd	$p_2 = 3 > 1 = p_4$ so interchange
$(4,2,5,1,3)$	2nd and 3rd	$p_2 = 3 = 3 = p_5$ so leave
$(4,2,5,1,3)$	1st and 2nd	$p_4 = 1 < 3 = p_2$ so leave
$(4,2,5,1,3)$		

position of the sequences after the four comparisons and appropriate inter-changes.

Next we pass through the current sequence again, interchanging jobs in the same fashion. Doing so guarantees that we will have moved the job with the second longest processing time into the $(n - 1)$st position of the sequence. Because we have already ensured that the last position is occupied by a job with the longest processing time, we need only compare and, perhaps, interchange the $(n - 2)$ pairs formed from the first $(n - 1)$ jobs in the sequence. These comparisons correspond with those in lines 5, 6 and 7 of the table. We continue this procedure, passing through the current sequence again and again, interchanging jobs so that during the rth pass the rth longest job is put into the $(n - r + 1)$st position. Ultimately we find an SPT sequence. Obviously, whenever two jobs have the same processing time their order is immaterial and, naturally, we do not interchange such jobs.

The pass which locates the rth longest job in the $(n - r + 1)$st position requires $(n - r)$ comparisons and at most the same number of inter-changes. Thus there will be

$$\sum_{r=1}^{n-1} (n - r) = \frac{n}{2} (n - 1)$$

comparisons and perhaps the same number of interchanges. Moreover, roughly the same amount of work will be required to increment and check the loop controls. Thus this algorithm will require $O(n(n - 1)/2) = O(n^2)$ operations to construct the SPT sequence. Any encoding convention will require at least one digit for each p_i. Hence n cannot exceed v, the instance size. It follows that the time complexity of the algorithm is at most $O(v^2)$ and, therefore, it is a polynomial time algorithm. (N.B. The sort procedure that we have used is by no means the most efficient available; see Aho $et\ al.$ (1974). However, our simple algorithm is all we need to show that a polynomial time algorithm exists.)

By similar arguments it may also be shown that all the constructive algorithms discussed earlier have polynomial time complexity. See Problem 9.6.6.

In Chapter 6 we determined that a dynamic programming solution to the $n/1/\bar{T}$ problem required $O(n2^{n-1})$ operations. This implies exponential time behaviour as we saw empirically in Table 6.5. It is an easy matter to confirm this behaviour theoretically. To specify an instance we need $2n$ numbers, viz, $p_1, p_2, \ldots, p_n$ and $d_1, d_2, \ldots, d_n$. We shall also need $2n - 1$ separators. Thus, if we construct an example in which each number may be represented by exactly k digits, we obtain an instance of size:

$$v = 2kn + 2n - 1$$

Solving for n, $$n = \frac{v + 1}{2(k + 1)}$$

Thus this instance will take

$$O\left(\frac{v + 1}{2(k + 1)} \cdot 2^{\left(\frac{v+1}{2(k+1)} - 1 \right)} \right) \tag{9.7}$$

operations to solve, which by definition implies exponential time complexity.

Note the difference between our deduction that an SPT schedule can be constructed in polynomial time and our deduction that dynamic programming requires exponential time. In the former we could use the simple inequality that $n < v$ and so deduce that all SPT schedules could be constructed in *at worst* $O(v^2)$ time. In the latter case, had we used the same approach we would have deduced that all dynamic programming solutions can have no worse time complexity behaviour than $O(v2^{v-1})$. But this tells us very little; $O(v)$ is no worse than $O(v2^{v-1})$. So the possibility of polynomial time behaviour would have remained. Thus we approached the deduction differently, showing by means of a particular instance that the behaviour could be *at best* that given by (9.7).

In conclusion we shall say a scheduling problem has been well solved if we have found an algorithm with polynomial time complexity. Such algorithms are usually constructive in nature and imply an understanding of the structure of the solution. On the other hand, dynamic programming and branch and bound solutions exhibit exponential time behaviour and do not constitute 'good' methods.

9.3 THE CLASSES P AND NP

We have now distinguished between two types of algorithm: those with polynomial time complexity and those without. This amounts to distinguishing two classes: P and NP. However, these are classes of problems, not classes of algorithms.

The class P consists of all problems for which algorithms with polynomial time behaviour have been found. The class NP is essentially the set of problems for which algorithms with exponential behaviour have been found. Clearly P is contained in NP; if we have a polynomial time algorithm for a problem, we can always inflate it inefficiently so that it takes exponential time. Also, occasionally a problem originally in NP but not in P is moved into P, as someone with a flash of insight discovers a polynomial time algorithm. Unfortunately, it is one of the most widely held conjectures of modern mathematics that there are problems in NP which may never be moved into P. In other words, many mathematicians believe that algorithms with polynomial time complexity will never be found for certain problems, essentially because they are just too hard. Most scheduling problems, alas, fall into this category.

Before discussing this conjecture, that $P \neq NP$, it will be advantageous to qualify the informal definition of these classes slightly. First we have implicitly suggested that the sort of problems in P and NP are **optimisation problems**; for example find an optimal schedule for an $n/m/G/\bar{F}$ job-shop problem. In fact, as computer scientists define P and NP, optimisation problems do not lie in these classes. Rather attention is confined to **decision or recognition problems.** These ask questions about existence that only require *yes* or *no* answers. For example, in a particular $n/m/G/\bar{F}$ instance does there exist a schedule with $\bar{F} \leq 37$? However, this restriction to recognition problems does not render the classes P and NP irrelevant to our concern with optimisation problems, because an optimisation problem has a polynomial time solution if and only if its associated recognition problem also has one. Consider the example above. If we can find an optimal schedule for the $n/m/G/\bar{F}$ problem, we can check whether the optimal mean flow time is less than 37, and so answer the recognition problem. If finding the optimal schedule takes polynomial time, then checking its value will not take much longer; so the recognition problem lies in P. Conversely, if the associated recognition problem is answerable in polynomial time, the optimisation problem is also. See Problem 6.6.10. Thus, in this sense an optimisation problem and its associated recognition problem are equally hard. In particular, if the recognition problem has so far defined attempts to find a polynomial time algorithm, the same must be true of the optimisation problem.

We have defined NP as the class of problems for which exponential time algorithms exist. For our informal purposes this is quite adequate. However, an alternative and better definition is that a recognition problem lies in NP if in polynomial time it is possible both to guess a feasible schedule and to check to see if this guess provides a *yes* answer. This definition does imply that exponential time algorithms exist for all problems in NP. To see this consider the following. For a particular problem suppose that for an instance of size v each guess and check takes less than $p(v)$ for some polynomial $p(v)$. There can be at most $v!$ possible schedules for an instance. (See Problem 9.6.3.) Thus, if we arrange to guess each possible schedule in turn, we shall exhaustively check all the possible schedules within $O(v!p(v))$ time. In short, complete enumeration of the problem takes exponential time.

Finally, it should be noted that the classes P and NP are not confined to scheduling, and the concepts apply to problems arising throughout combinatorial theory. See Garey and Johnson (1979), and Klee (1980).

Having thus qualified the definition of P and NP, we return to our discussion of the conjecture that P $\neq$ NP and its implications for scheduling problems. A central concept in the argument will be the idea of **reducing** one problem to another. To introduce this through an example, consider the discussion in Chapter 8. There it was shown that a certain scheduling problem could be translated into an equivalent integer programming problem. In the terminology now to be adopted, we reduced our scheduling problem to an integer programme. Admittedly those were optimisation problems, whereas now our concern is with recognition problems; but the underlying idea of reducibility is the same.

Suppose that we can reduce one problem Π_1 to another Π_2 in polynomial time; then we say Π_1 is **polynomially reducible** to problem Π_2, and we write

$$\Pi_1 \propto \Pi_2,$$

Explicitly this means that, given any instance of Π_1 with size v we can construct an instance of Π_2 within $p(v)$ operations for some polynomial $p(v)$. Moreover, the two instances are equivalent in the sense that the first has a *yes* answer if and only if the second does. We also note that the size of the constructed instance of Π_2 must be less than $p(v)$, because to construct each symbol in its list must take at least one operation.†

Suppose that $\Pi_1 \propto \Pi_2$ and that Π_2 lies in P. Then we can easily construct a polynomial time algorithm for answering Π_1, demonstrating that it to lies in P. Quite simply, given an instance of Π_1 of size v we can reduce in time bounded by a polynomial $p(v)$ to an instance of Π_2 of size at most $p(v)$. Because Π_2 lies in P, any instance of Π_2 of size η can be answered in polynomial time, say $q(\eta)$. Thus we can answer our equivalent instance in $q(p(v))$. Adding the time for reduction, we can answer the original Π_1 instance in $p(v) + q(p(v))$, that is in polynomial time. Conversely, if $\Pi_1 \propto \Pi_2$ and if, at present, Π_1 does not lie in P, then neither can Π_2. Thus the relation of polynomial reducibility between problem types may be viewed as a 'hardness' ordering of problems. If $\Pi_1 \propto \Pi_2$, then problems of type Π_2 must be at least as hard to solve as problems of type Π_1. This ordering of problems according to hardness is so crucial to the following arguments that it is of the greatest importance to understand it fully.

It is easy to show (see Problem 9.6.8) that if Π_1, Π_2 and Π_3 are arbitrary problems then

$$(\Pi_1 \propto \Pi_2 \quad \text{and} \quad \Pi_2 \propto \Pi_3) \Rightarrow \Pi_1 \propto \Pi_3.$$

Because of this we say $\propto$ is **transitive**.

---

†Strictly, this result is not quite a logical consequence of our informal definitions. To justify it completely requires a much deeper treatment of the time complexity of algorithms, such as given by Garey and Johnson (1979). However, since the result is so intuitive I have no qualms in asking you to accept it.

We say that a problem Π lying in NP is **NP-complete** if every other problem in NP is polynomially reducible to Π, that is

$$\Pi' \propto \Pi \quad \text{for all } \Pi' \text{ lying in NP.} \tag{9.8}$$

Thus the NP-complete problems form a subclass of NP. Furthermore, this subclass is formed of the hardest problems in NP. For, if we find a polynomial time algorithm for any NP-complete problem, then we can answer *all* problems in NP in polynomial time. Thus if one such algorithm is found, $P = NP$. The conjecture that $P \neq NP$ is, therefore, equivalent to the conjecture that no NP-complete problem can be answered in polynomial time.

Obviously it is not feasible to present a completely convincing argument that this conjecture is true. To do so would be to prove the conjecture. Nonetheless, I can present some persuasive empirical evidence can be found. Although the concept of NP-completeness was introduced only recently (Cook, 1971; Karp, 1972), for very many years people have been trying to solve problems that have been shown subsequently to be NP-complete. Garey and Johnson (1979) list about 300 different NP-complete problem types. Each of these has been studied extensively by some of the greatest minds of our time, but none has had the insight to devise a polynomial time algorithm. In fact, most people who have worked in the area have become convinced that NP-complete problems cannot be solved in polynomial time. This conviction arises not just because it is hard to admit failure, but familiarity with NP-complete problems brings a deep appreciation of their very great difficulty—one does not know how hard a problem is until one has tried it. (Hence in Chapter 1 I insisted that *you* tried to schedule Algy and friends' reading, and not leave the effort just to me.) Despite all this empirical evidence, it cannot be repeated too often, that $P \neq NP$ remains a conjecture.

And now to the point of this chapter. It has been shown that very many scheduling problems are NP-complete. For instance, both the $n/1//\Sigma \gamma_i$ and the general $n/3/F/C_{max}$ problems are NP-complete. Hence, for the present there is no alternative to trying to solve these by implicit enumeration, as we did in Chapters 6 and 7. The introduction of non-zero ready times usually renders a problem NP-complete; thus the $n/1//\overline{C}$, $n/1//L_{max}$, and $n/1//n_T$ problems all became NP-complete when non-zero ready times are allowed. The $n/m/F/C_{max}$ problem is NP-complete for all $m \geq 3$ and so naturally is the $n/m/G/C_{max}$ problem. Thus it is not at all surprising that constructive algorithms have not been found for these problems. To have done so would have resolved one of the great conjectures of our time.

Strictly, of course, we should remember that the concept of NP-completeness applies to recognition problems. Thus, in saying that a scheduling problem is NP-complete, we are referring to the recognition problem associated with the optimisation problem. From our earlier discussion we know that polynomial time algorithms exist either for both problems or for neither, but not for one alone. Hence, when the associated recognition problem is NP-complete, we cannot solve the optimisation problem in polynomial time and we say that it is **NP-hard**. However, given the informality of our treatment the distinction between NP-completeness and NP-hardness is somewhat pedantic, and we shall often allow it to blur.

We complete the section with a discussion of **NP-completeness in the strong sense** or **unary NP-completeness**. This concept will not be needed until Section 11.3, but it is convenient to introduce it within the context of our general treatment of NP-completeness. Remember that in defining the size of an instance we have limited the choice of encoding convention to those which encode numbers to bases other than unity. This, we have seen, means that the size depends linearly on the number of data, but only logarithmically on the magnitude of the numbers involved. Perhaps, if we were to allow the instance size to grow linearly with the magnitudes, we might be able to find polynomial time algorithms with respect to this definition of size. If we knew that the data were always small, then such an algorithm would answer this problem *for all practical purposes*. It is this idea that we try to encapsulate. In fact, we shall not do this by encoding numbers to base 1. Rather, we retain our original definition of instance size and limit the magnitude of the data by including an upper bound in the time complexity function.

So, returning to the original definition of instance size with all numbers encoded to some base other than 1, we shall say that an algorithm has **pseudo-polynomial time complexity** if its maximum number of operations is bounded by a polynomial $q(v, \lambda)$, where v is the instance size and λ is an upper bound on the magnitude of each of the data. In essence, we say a problem Π is NP-complete in the strong sense if it cannot be answered with a pseudo-polynomial time algorithm. However, it is very difficult to prove that a problem satisfies this non-existence property. (In fact such a proof would also show $P \neq NP$.) Thus the formal definition uses the concept of NP-completeness already defined. For a recognition problem Π and a polynomial $p(v)$ of the problem size, we define the sub-problem Π_p with the restriction that all data are bounded by $p(v)$. Thus, if we have a pseudo-polynomial time algorithm for Π with complexity $q(v, \lambda)$ as above, Π_p is answerable in $q(v, p(v))$ operations, that is polynomial time. Hence Π_p lies in P. We say that Π is **NP-complete in the strong sense** if for some polynomial $p(v)$, Π_p is NP-complete. Thus, assuming that $P \neq NP$, Π cannot be answered in pseudo-polynomial time if it is NP-complete in the strong sense.

We made a distinction between the NP-completeness of a recognition problem and the NP-hardness of an optimisation problem. In the same way we shall make a distinction here, although we shall come to no great harm if we allow it to blur. An optimisation problem will be called **NP-hard in the strong sense**, if the recognition problem derived from it is NP-complete in the strong sense.

9.4 THE NP-COMPLETENESS OF $n/1//L_{max}$ WITH NON-ZERO READY TIMES

It would be out of place in this introductory treatment to spend much time on proofs of NP-completeness. Generally such proofs are very involved and, more importantly, require a wide knowledge of problems whose NP-completeness has already been established. Nonetheless, it is appropriate to study one so that you may be introduced to the underlying ideas.

Cook (1971) was the first person to prove that a particular problem was NP-complete. He showed from first principles that every problem in NP can be polynomially reduced to an archetypal recognition problem called **satisfiability**. Thus satisfiability is NP-complete.

When at least one NP-complete problem is known it becomes easier, relatively speaking, to prove that other problems lie in this class. Suppose that we wish to show Π is NP-complete. Our aim will be to find a known NP-complete problem Π' such that $\Pi' \propto \Pi$. Suppose that we find such a Π'. Since it is NP-complete, $\Pi'' \propto \Pi'$ for all problems Π'' in NP. Since $\propto$ is transitive and $\Pi' \propto \Pi$, we must also have that $\Pi'' \propto \Pi$ for all problems Π'' in NP. Hence Π is also NP-complete. The skill in these proofs comes in selecting an appropriate Π' and then devising a suitable polynomial reduction. Garey and Johnson (1979) give some very useful general advice.

As an example of these proofs we select one of the more straightforward. Following a demonstration given by Rinnooy Kan (1976), we show that $n/1//L_{max}$ problem with non-zero ready times is NP-complete. Remember that in Chapter 3 we proved that the EDD sequence, which may be constructed in polynomial time, solves the $N/1//L_{max}$ problem with zero ready times. Thus it is the introduction of non-zero ready times that leads to the difficulty.

Our method will be to show that a known NP-complete problem called the **knapsack problem** is polynomially reducible to this scheduling problem. Since our familiarity with recognition problems is rather limited, we explicitly state both knapsack and $n/1//L_{max}$ in this form.

Knapsack. Given a set of N positive numbers $A = \{a_1, a_2, \ldots, a_N\}$ and a positive number $b \leqslant \sum_{i=1}^{N} a_i$, is there a subset S of A such that

$$\sum_{a_i \, in \, S} a_i = b?$$

$n/1//L_{max}$. Given an n-job, 1-machine problem with processing time p_i, ready times r_i, due dates d_i, $i = 1, 2, \ldots, n$, and a number y, is there a schedule such that $L_{max} \leqslant y$?

Our first task in proving the NP-completeness of $n/1//L_{max}$ is to show it lies in NP. But this is obviously true. To check if a guessed schedule has $L_{max} \leqslant y$ is very easy and can clearly be accomplished in a number of operations bounded by a polynomial in n. We have often remarked that $n \leqslant v$ the problem size. So $n/1//L_{max}$ lies in NP.

Our next aim is to show that in polynomial time we can turn any knapsack problem into an equivalent $n/1//L_{max}$ problem such that the former has a *yes* answer if and only if the latter does. Consider the following construction.

Given the knapsack problem above, define an $n/1//L_{max}$ problem by:

$$n = N + 1;$$

$$r_i = 0, p_i = a_i, d_i = \left(\sum_{i=1}^{N} a_i\right) + 1 \quad \text{for} \quad i = 1, 2, \ldots, N;$$

$$r_{N+1} = b, p_{N+1} = 1, d_{N+1} = b + 1;$$

$$y = 0.$$

Note that the only non-zero ready time is for J_{N+1}. It is obvious that this construction of an $n/1//L_{max}$ problem from a knapsack problem can be accomplished in polynomial time. We only have to assign $(3N + 5)$ easily calculated numbers. We will show that this $n/1//L_{max}$ problem can have $L_{max} \leqslant 0$ if and only if we can find a subset S of $\{a_1, a_2 \ldots a_N\}$ such that $\sum_{a_i \text{ in } S} a_i = b$.

Suppose first that such a subset S exists. Consider a schedule given by first processing the jobs corresponding to S in any order then the job J_{N+1}, and finally the remaining jobs in any order. See Fig. 9.1. The jobs in S complete processing at

$$\sum_{a_i \text{ in } S} p_i = \sum_{a_i \text{ in } S} a_i = b, \text{ by assumption.}$$

The job J_{N+1} completes at $b + 1$. The remaining jobs are then processed without idle time and complete at $\sum_{i=1}^{N} a_i + 1$. Therefore no job under this schedule is tardy. So

$$L_{max} \leqslant 0 = y.$$

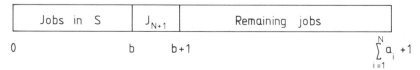

Fig. 9.1 The schedule when a subset S exists such that $\sum_{a_i \text{ in } S} a_i = b$.

Thus if knapsack has a *yes* answer so does $n/1//L_{max}$.

Suppose now that knapsack has a *no* answer. Then there is no subset S of $\{a_1 a_2 \ldots a_N\}$ such that $\Sigma_{a_i \text{in} S} a_i = b$. Consider any schedule for the $n/1//L_{max}$ problem. Let S be the set of jobs processed before J_{N+1}. Then these jobs are completed at

$$C_s = \Sigma_{J_i \text{in} S} p_i = \Sigma_{J_i \text{in} S} a_i \neq b, \text{ by assumption.}$$

If $C_S > b$, then J_{N+1} may start immediately and will complete at $C_S + 1 > b + 1$. So J_{N+1} will be tardy, i.e. $L_{N+1} > 0$, and hence $L_{max} > 0 = y$. If $C_S < b$, then J_{N+1} cannot start processing immediately because $r_{N+1} = b$. Thus there must be some idle time. As we shall see, this idle time means that at least one of the jobs scheduled after J_{N+1} must be late. (N.B. Since we assume in knapsack that $b \leqslant \Sigma_{i=1}^N a_i$, $C_S < b \Rightarrow$ at least one job must be scheduled after J_{N+1}.) The processing of the remaining jobs will take a total time $(\Sigma_{i=i}^N a_i) - C_S$. So

$$C_{max} = b + 1 + \left(\sum_{i=1}^N a_i\right) - C_S$$

$$> \sum_{i=1}^N a_i + 1, \quad \text{since} \quad C_S < b.$$

Because each of these remaining jobs is due at $\Sigma_{i=1}^N a_i + 1$, at least one must be late. So $L_{max} > 0 = y$. Thus if knapsack has a *no* answer so does the constructed scheduling problem.

So knapsack and $n/1//L_{max}$ are equivalent problems as required, and we may deduce that $n/1//L_{max}$ is NP-complete.

9.5 REFERENCES AND FURTHER READING

The central reference for this chapter has been Garey and Johnson (1979); indeed, currently it is the definitive reference on computational complexity. Ullman (1976) gives a treatment of NP-completeness aimed specifically at an audience whose main concern, like ours, is with sequencing and scheduling. Lenstra (1977) and Rinnooy Kan (1976) both indicate the importance of considering the computational complexity of scheduling problems, but neither introduce the topic from first principles. See also Lenstra and Rinnooy Kan (1979).

There are many papers that demonstrate that particular scheduling problems are NP-complete; the majority of these are surveyed by Rinnooy Kan (1976), Lenstra *et al.* (1977), and Graham *et al.* (1979). Garey *et al.* (1976) showed that a number of important problems are NP-complete: namely, $n/m/F/C_{max}$ with $m \geqslant 3$, $n/m/G/\overline{F}$ with $m \geqslant 2$ and repeated processing on the same machine allowed.

We remarked in Section 4.1 and showed in Section 6.4 that the introduction of precedence constraints can actually make problems easier to solve because the number of feasible solutions may be reduced substantially. This does not mean that their presence allows constructive algorithms to be found for problems that are NP-complete in their absence. Rather it means that enumeration algorithms have less dramatic exponential time complexity when precedence constraints are present than when they are not. Indeed, in terms of computational complexity our statement that problems become easier is somewhat dubious; some problems become NP-complete upon the introduction of precedence constraints. Lenstra and Rinnooy Kan (1978) discuss this behaviour in some detail. See also Section 12.2.

In this chapter we have used polynomial reducibility purely as an intermediate concept; our ultimate objective was to define the class of NP-complete problems. Recently Monma (1980) has used the polynomial reducibility of one problem to another in a much more positive manner. Working within the class P he has developed a general approach to permutation scheduling problems when the objective is to minimise the maximum cost of a job. Essentially, he polynomially reduces problems to an archetypal maximum cost problem for which a polynomial time algorithm is known. In doing so he provides a unified approach to Lawler's Algorithm (Section 4.3) and Johnson's Algorithm (Section 5.3), as well as several other lesser known problems.

9.6 PROBLEMS

1. Show that, if v is the size of a problem when all the data are represented in binary and η is its size when the same listing convention is adopted but numbers are represented to the base 10, then

$$v \leq (\log_2 10)\eta,$$

and $$\eta \leq v.$$

2. Explaining your conventions clearly, determine the size of the problem given in Problems 3.7.8 and 6.6.1.

3. Show that if an $n/m/A/B$ problem has size v then

$$v! \geq (n!)^m.$$

(*Hint*: show $v \geq n.m$).

4. Show that if $f(v)$ is $O(v^n)$ and $g(v)$ is $O(v^m)$ then

(i) $f(v)g(v)$ is $O(v^{n+m})$;
(ii) $f(v)/g(v)$ is $O(v^{n-m})$;
(iii) $f(g(v))$ is $O(v^{nm})$.

5. Show that the polynomial $a_n v^n + a_{n-1}v^{n-1} + \ldots + a_0$ is $O(v^n)$.

6. Show that the following algorithms have polynomial time complexity

(i) Johnson's for the $n/2/F/F_{max}$ problem,

(ii) Lawler's for the $n/1//max_{i=1}^{n}\{\gamma_i(C_i)\}$ problem

7. Suppose that a problem has size v under one reasonable convention and size η under another, where

$$\eta \leqslant p(v) \quad \text{and} \quad v \leqslant q(\eta)$$

for some polynomials $p(v)$ and $q(\eta)$. Show that, if a solution algorithm has polynomial time complexity with respect to v, it also has polynomial time complexity with respect to η.

8. Prove that $\propto$ is transitive.

9. (To understand and solve this question you must have studied equivalence relations elsewhere.)
Define the relation $\sim$ between problem types by

$$\Pi_1 \sim \Pi_2 \Leftrightarrow (\Pi_1 \propto \Pi_2 \quad \text{and} \quad \Pi_2 \propto \Pi_1.)$$

Show that $\sim$ is an equivalence relation and deduce that P is an equivalence class of NP under this relation. Also show that the class of NP-complete problems is an equivalence class.

10. Suppose that we have polynomial time algorithm for answering the recognition problem: given an $n/m/G/\bar{F}$ instance, does there exist a schedule with $\bar{F} \leqslant y$? Show that the $n/m/G/\bar{F}$ problem is solvable in polynomial time.
Hint: Use an encoding convention which represents the data in binary form. Show that for all schedules in an instance, $\bar{F} \leqslant 2^v$. Use the recognition algorithm to locate the optimal $\bar{F}$ in intervals of the form $(k/2^{m-1}) \leqslant \bar{F} \leqslant ((k + 1)/2^{m-1})$. Hence find the optimal value of $\bar{F}$ in at most $[v + \log_2 v + 1]$ steps.

Chapter **10**

Heuristic Methods: General Approaches

10.1 INTRODUCTION

It is worth pausing to put the pessimism of the previous chapter into perspective. For problems that are NP-hard, which includes most arising in scheduling, there are at present no easy solutions. Furthermore, if informed mathematical opinion is correct, there never will be any easy solutions. The only methods available are those of implicit (or explicit) enumeration, which may take a prohibitive amount of computation. Certainly large NP-hard scheduling problems are for all practical purposes insoluble.

Stated in this way, the theory of NP-completeness gives us a very negative outlook on scheduling. The implication is that we should test each problem that arises to find whether it is NP-hard; or rather we should search the literature to see whether it has already been shown to be NP-hard—there is no point in duplicating effort. If we determine that it is a large NP-hard problem, then apparently we should acknowledge defeat, for it is probably impossible to find an optimal solution within a human lifetime. But in practice we cannot simply leave the problem unsolved. Scheduling problems are not just intellectual exercises invented for the pleasure of mathematicians. They have their basis in reality; they arise because the processing of jobs does have to be sequenced. Of necessity some schedule must be used, and we know from experience and from theory that the cost of processing may depend crucially upon that choice of schedule. If we cannot find the best schedule for the problem within a reasonable time, then surely we should not abandon all analysis and pick at random. Instead we should use our knowledge and experience to find a schedule which, if not optimal, may at least be expected to perform better than average. Here we consider algorithms to do just this.

They are called **heuristic or approximation algorithms**. If anything, the former term refers to methods justified purely because they seem sensible, that is by 'hand-waving' arguments, and because empirically they are found to perform 'fairly well'. The latter term is occasionally reserved for algorithms that produce solutions guaranteed to be within a certain distance of the optimum (see next chapter). But, by and large, the terms are used interchangeably and we shall make no strong distinction here.

Thus NP-completeness should not be seen as simply providing a pessimistic outlook on our subject. Its main import is that it provides us with a test of the hardness of problems. If a problem is large and NP-hard, then we must consider using heuristic methods. However, we should emphasise that the NP-hardness of a problem alone is not sufficient reason to resort to heuristic methods. It must also be so large that enumerative methods are intractable. Thus:

Assumptions for Chapter 10
We make no assumptions to limit the class of problems studied in this chapter, but we do make one strong assumption about the use of heuristic methods. We assume that they are not used when either a constructive, polynomial time solution exists or implicit enumeration is computationally feasible. One should not accept approximations when optimal solutions may be found just as easily.

Some heuristic methods are applicable to wide classes of problem being essentially approaches to problem-solving rather than problem-specific algorithms. Others are *ad hoc* 'rules of thumb' only applicable to the very specific class of problems for which they were designed. The organisation of this chapter is such that the more generally applicable methods come first.

10.2 SCHEDULE GENERATION TECHNIQUES

Our study begins with a family of methods that spans the spectrum between complete enumeration and heuristic solution. We consider algorithms that are capable of producing all, some or just one schedule in a particular class. If that class is sure to contain an optimal schedule and if we generate, perhaps implicitly, all that class, then we are again in the realm of optimal solution by enumeration. If, on the other hand, either the class is not guaranteed to contain an optimal schedule or we do not generate all the class, then we may only find an approximate solution. All these possibilities will be discussed and clarified in the following. The setting will be that of the general job shop: $n/m/G/B$.

In Section 2.2 we discussed semi-active timetabling. Semi-active schedules ensure that each operation is started as soon as it can be, while obeying both the technological constraints and the processing sequence. Here we consider two further classes of schedules: the active and the non-delay.

In an **active schedule** the processing sequence is such that no operation can be started any earlier without either delaying some other operation or violating the technological constraints. The active schedules form a sub-class of the semi-active, i.e. an active schedule is necessarily semi-active. The distinction between active and semi-active schedules may be appreciated by looking back to Chapter 1. Figures 1.1, 1.3 and 1.4 give different schedules for our newspaper-reading friends. All three are semi-active, as we have noted before. However, the schedule in Fig. 1.1 is not active. Consider the *Financial Times*. Both Charles and Bertie could be given the paper before Digby and not cause him any delay; for they would finish it by 11.00 a.m. In Fig. 1.3 this has been done and you may check that this is an active schedule. No paper can be given any earlier to anyone without delaying someone else or violating their desired reading order, i.e. the technological constraints. In semi-active timetabling we put the blocks representing operations onto the Gantt diagram and *slide* them to the left as far as we can. In active timetabling we also allow ourselves the possibility of *leap frogging* one block over another, provided that there is sufficient idle time waiting to receive it. (Of course, this leap-frogging will change the processing sequence, which is given to the timetabling procedure.)

In a **non-delay schedule** no machine is kept idle when it could start processing some operation. These schedules form a sub-class of the active. (See Problem 10.7.2.) So a non-delay schedule is necessarily active and, hence, necessarily semi-active. Consider the schedule shown in Fig. 1.3. This is a non-delay schedule because no paper lies unread when there is someone free to read it.

In Theorem 2.1 we showed that there is necessarily an optimal schedule, which is semi-active. It is easy to modify the proof of this theorem to show that there is necessarily an optimum, which is active. (See Problem 10.7.1.) Thus we may without loss restrict our attention to the class of active schedules. Since this class is smaller than that of the semi-active, we make our task of finding an optimal schedule earlier by this restriction. The class of non-delay schedules is smaller still. However, there need not be an optimal schedule, which is non-delay. This may be seen by considering Fig. 1.4, which shows the *unique* optimum. (You may check that this is the unique optimum by extending the argument given in Section 1.8.) This schedule is certainly *not* non-delay. Remember it achieves a final completion time of 11.30 precisely because it insists on Algy's and Bertie's patience; both the *Financial Times* and the *Guardian* could be being read before 9.05 and 8.50 respectively. Nonetheless, although there is not necessarily an optimal schedule in the class of non-delay ones, we shall encounter strong empirical reasons for designing heuristic algorithms that only generate non-delay schedules.

We next discuss an algorithm to generate some or all of the active schedules for a problem. Shortly we shall modify it to generate non-delay schedules. Suppose that we have the data for an $n/m/G/B$ problem and we want to generate an active schedule. We could simply work through a series of semi-active schedules testing each until we found an active one. Need it be said that this would be a very time-consuming procedure? Fortunately we have a method due to Giffler and Thompson (1960) which constructs active schedules *ab initio*. See also Heller and Logemann (1962).

To describe their algorithm we need some more notation and terminology. In the algorithm we shall schedule operations one at a time. We shall say that an operation is **schedulable** if all those operations which must precede it within its job have already been scheduled. Since there are nm operations, the algorithm will iterate through nm stages. At stage t let

P_t – be the partial schedule of the $(t - 1)$ scheduled operations;
S_t – be the set of operations schedulable at stage t, i.e. all the operations that must precede those in S_t are in P_t;
σ_k – be the earliest time that operation o_k in S_t could be started;
ϕ_k – be the earliest time that operation o_k in S_t could be finished, that is $\phi_k = \sigma_k + p_k$, where p_k is the processing time of o_k.

N.B. According to the notation of Section 1.3, operations have 2 subscripts indicating their association with both a job and a machine. Here, however, we shall not be concerned explicitly with the associated job and machine so a single subscript is adopted for clarity.

Algorithm 10.1 (Giffler and Thompson)

Step 1 Let $t = 1$ with P_1 being null. S_1 will be the set of all operations with no predecessors; in other words, those that are first in their job.
Step 2 Find $\phi^* = \min_{o_k \text{ in } S_k} \{\phi_k\}$ and the machine M^* on which ϕ^* occurs. If there is a choice for M^*, choose arbitrarily.
Step 3 Choose an operation o_j in S_t such that
      (1) it requires M^*, and
      (2) $\sigma_j < \phi^*$.
Step 4 Move to next stage by
      (1) adding o_j to P_t, so creating P_{t+1};
      (2) deleting o_j from S_t and creating S_{t+1} by adding to S_t the operation that directly follows o_j in its job (unless o_j completes its job);
      (3) incrementing t by 1.
Step 5 If there are any operations left unscheduled ($t < nm$), go to *Step 2*. Otherwise, stop.

First let us check that this algorithm produces an active schedule. To remind you, that is one in which no operation could be started earlier without delaying some other operation or breaking the technological constraints. At each stage the next operation to schedule is chosen from S_t. So the algorithm can never break the technological constraints. Hence to prove that an active schedule is generated we must show that to start any operation earlier must delay the start of some other. With this in mind let us work through the algorithm.

Step 1 simply initialises t, P_1 and S_1 in the obvious fashion. So suppose that we come to step 2 at stage t with $0 \leq t < mn$ and with P_t and S_t set appropriately by the earlier stages. Consider ϕ^*. It gives the earliest possible completion time of the next operation added to the partial schedule P_t. Next consider the machine M^* on which ϕ^* is attained. (We may choose M^* arbitrarily if there is a choice; see Problem 10.8.3.) In step 2, we examine all schedulable operations that need M^* and can start before ϕ^*. There is at least one such operation, namely one that completes at ϕ^*. Of these we select one and schedule it to start as soon as it can. The method of selection here will be the subject of much discussion shortly. We have selected an operation o_j for M^* that can start strictly before ϕ^*. Moreover, if we try to start any of the other operations schedulable on M^* *before* the selected operation o_j, we know that they must complete at some time $\geq \phi^*$. Hence they must delay the start of o_j. We see, therefore, that this algorithm is steadily building up an active schedule. Step 4 simply moves from one stage to the next in the obvious fashion, while step 5 determines whether the complete schedule has been generated.

It should be noted that for any active schedule there is a sequence of choices at step 3 which will lead to its generation. See Problem 10.7.4.

As an example of the use of the algorithm we shall generate an active schedule for the newspaper reading problem of Chapter 1. In doing so we shall use the notation (A,S) to represent Algy's reading of the *Sun*, and similarly for the other operations. For convenience the data are repeated in Table 10.1. Table 10.2 gives the calculations needed to generate an active schedule. At each stage any choice at step 3 of the algorithm has been made arbitrarily. The calculations in the table are straightforward. It should be noted that, the scheduling of an operation in step 4 not only has implications for the earliest moment when the paper concerned is free at the next stage, but also will change the earliest start and finish times of other operations requiring that paper. Thus, when Algy is given the *Guardian* at stage 4, the earliest time that Bertie can begin to read it drops from 9.05 to 10.00. The resulting schedule is shown in the Gantt chart of Fig. 10.1.

(N.B. When modifying the layout of Table 10.2 for other scheduling problems it should be remembered that there the papers are the machines and the readers are the jobs.)

Table 10.1—The data for the newspaper-reading problem

Reader	Gets up at	Reading Order and Times in minutes			
Algy	8.30	FT	G	DE	S
		(60)	(30)	(2)	(5)
Bertie	8.45	G	DE	FT	S
		(75)	(3)	(25)	(10)
Charles	8.45	DE	G	FT	S
		(5)	(15)	(10)	(30)
Digby	9.30	S	FT	G	DE
		(90)	(1)	(1)	(1)

Table 10.2—Generation of an active schedule

Stage t	Paper next free at:				o_k in S_t	σ_k	ϕ_k	ϕ^*	operation scheduled o_j
	FT	G	DE	S					
0	0	0	0	0	(A, FT)	8.30	9.30		
					(B, G)	8.45	10.00		
					(C, DE)	8.45	8.50	8.50	(C, DE)
					(D, S)	9.30	11.00		
1	0	0	8.50	0	(A, FT)	8.30	9.30		
					(B, G)	8.45	10.00		
					(D, S)	9.30	11.00		
					(C, G)	8.50	9.05	9.05	(C, G)
2	0	9.05	8.50	0	(A, FT)	8.30	9.30		(A, FT)
					(B, G)	9.05	10.20		
					(D, S)	9.30	11.00		
					(C, FT)	9.05	9.15	9.15	
3	9.30	9.05	8.50	0	(B, G)	9.05	10.20		
					(D, S)	9.30	11.00		
					(C, FT)	9.30	9.40	9.40	(C, FT)
					(A, G)	9.30	10.00		
4	9.40	9.05	8.50	0	(B, G)	9.05	10.20		
					(D, S)	9.30	11.00		
					(A, G)	9.30	10.00	10.00	(A, G)
					(C, S)	9.40	10.10		

Table 10.2—*continued*

Stage t	Paper next free at:				o_k in S_t	σ_k	ϕ_k	ϕ^*	operation scheduled o_j
	FT	G	DE	S					
5	9.40	10.00	8.50	0	(B,G)	10.00	11.15		
					(D,S)	9.30	11.00		
					(C,S)	9.40	10.10		
					(A,DE)	10.00	10.02	10.02	(A,DE)
6	9.40	10.00	10.02	0	(B,G)	10.00	11.15		
					(D,S)	9.30	11.00		
					(C,S)	9.40	10.10		
					(A,S)	10.02	10.07	10.07	(A,S)
7	9.40	10.00	10.02	10.07	(B,G)	10.00	11.15		
					(D,S)	10.07	11.37		(D,S)
					(C,S)	10.07	10.37	10.37	
8	9.40	10.00	10.02	11.37	(B,G)	10.00	11.15	11.15	(B,G)
					(C,S)	11.37	12.07		
					(D,FT)	11.37	11.38		
9	9.40	11.15	10.02	11.37	(C,S)	11.37	12.07		
					(D,FT)	11.37	11.38		
					(B,DE)	11.15	11.18	11.18	(B,DE)
10	9.40	11.15	11.18	11.37	(C,S)	11.37	12.07		
					(D,FT)	11.37	11.38	11.38	
					(B,FT)	11.18	11.43		(B,FT)
11	11.43	11.15	11.18	11.37	(C,S)	11.37	12.07		
					(D,FT)	11.43	11.44	11.44	(D,FT)
					(B,S)	11.43	11.53		
12	11.44	11.15	11.18	11.37	(C,S)	11.37	12.07		
					(B,S)	11.43	11.53		
					(D,G)	11.44	11.45	11.45	(D,G)
13	11.44	11.45	11.18	11.37	(C,S)	11.37	12.07		
					(B,S)	11.43	11.53		
					(D,DE)	11.45	11.46	11.46	(D,DE)
14	11.44	11.45	11.46	11.37	(C,S)	11.37	12.07		(C,S)
					(B,S)	11.43	11.53	11.53	
15	11.44	11.45	11.46	12.07	(B,S)	12.07	12.17		(B,S)

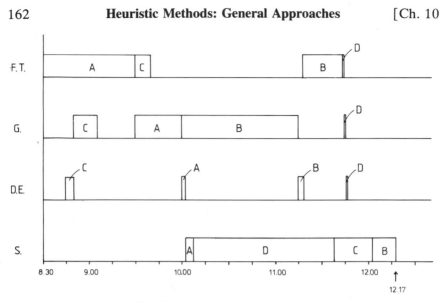

Fig. 10.1 The generated active schedule.

It is an easy matter to modify this algorithm so that it produces non-delay schedules:

Algorithm 10.2

Step 1 Let $t = 1$, P_1 being null. S_1 will be the set of all operations with no predecessors, in other words, those that are first in their job.

Step 2 Find $\sigma^* = \min_{o_k \text{ in } S_k}\{\sigma_k\}$ and the machine M^* on which σ^* occurs. If there is a choice for M^*, choose arbitrarily.

Step 3 Choose an operation o_j in S_t such that
  (1) it requires M^*, and
  (2) $\sigma_j = \sigma^*$.

Step 4 Move to next stage by
  (1) adding o_j to P_t so creating P_{t+1};
  (2) deleting o_j from S_t and creating S_{t+1} by adding to S_t the operation that directly follows o_j in its job (unless o_j completes its job);
  (3) incrementing t by 1.

Step 5 If there are any operations left unscheduled ($t \leqslant mn$), go to *step 2*. Otherwise, stop.

Problem 10.7.5 asks you to confirm that this algorithm always does produce a non-delay schedule and, moreover, that any non-delay schedule can be generated in this way.

   So we have two algorithms, the first for generating active schedules, the second for non-delay. However, neither is fully defined; there is an undefined choice at step 3. To this we now turn.

First we may pass through either algorithm many times making all possible sequences of choices at step 3. Thus we would generate all the active or all the non-delay schedules. Since the set of active schedules is guaranteed to contain an optimum, generating all and selecting the best leads to an optimal solution. There are often far fewer active schedules than semi-active; so this process of complete enumeration may be feasible for small problems. Nonetheless, as the theory of NP-completeness predicts, such an approach rapidly becomes computationally infeasible as the problem size grows. Some respite may be gained by sacrificing optimality and only generating non-delay schedules. There are fewer of these generally and empirically it is known that the best non-delay schedule, if not optimal, is certainly very good. But even this is infeasible for problems of a moderate size.

An obvious progression from complete enumeration is to consider implicit enumeration. Thus these algorithms may be used as a basis for a branch and bound solution. In our study of branch and bound the elimination tree for the $n/3/F/C_{max}$ example enumerated all possible permutation schedules by making all possible sequences of choices of the job to place 1st, 2nd, 3rd, etc. Here we again have to enumerate all possible sequences of choices, but each choice now concerns the operation to schedule at step 3 in the generation algorithm. Thus a similar branching structure develops. See Fig. 10.2. If we can find computationally feasible lower bounds, we may develop a full branch and bound solution. When this is based upon active schedule generation, an optimal schedule is found. Again it may be worth sacrificing optimality and gaining computationally by searching the generally smaller tree associated with non-delay schedules. However,

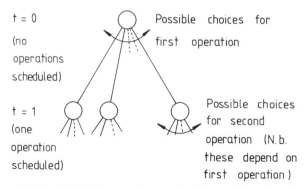

Fig. 10.2 The elimination tree for an $n/m/G/B$ problem based upon schedule generation.

although suggested in the literature (Baker, 1974), few seem to have done this. The early work on branch and bound in this area was done by Brooks and White (1966). For more recent work see the references given in Section 7.5.

However, even with the best bounding schemes available it soon becomes computationally infeasible to use these methods (Lageweg *et al.*, 1977). So for most problems we cannot avoid the problem of making the choice at step 3 simply by making all possible choices. We must be more selective. Many **selection or priority rules** have been discussed in the literature. Panwalker and Iskander (1977) list over 100. Of these the six most studied are the following.

SPT   (Shortest Processing Time) Select an operation with the shortest processing time.

FCFS   (First Come First Served) Select an operation that has been in S_t for the greatest number of stages.

MWKR   (Most Work Remaining) Select an operation that belongs to the job with the greatest total processing time remaining.

LWKR   (Least Work Remaining) Select an operation that belongs to the job with the least total processing time remaining.

MOPNR   (Most Operations Remaining) Select an operation which belongs to the job with the greatest number of operations still to be processed.

RANDOM (Random) Select an operation at random.
(N.B. This random choice is easy to program on computers since they have random number generation routines.)

The reasons for suggesting these rules are not hard to see. Using the SPT rule finds an operation which monopolises the machine M^* for the least time and, hence, perhaps constrains the whole system least. The FCFS rule is based upon a notion of fairness and is perhaps more applicable in queueing theory, from which it is derived. The MWKR rule is based upon the notion that it may be best not to get too far behind on a particular job; whereas the LWKR rule stems from the contrary notion that perhaps one should concentrate on finishing some jobs first in the hope that the later jobs may be speeded through a less cluttered system. Finally, the feeling that delays in scheduling derive from the change-over of jobs between machines suggests that we should hurry through the jobs with the most changeovers left, hence the MOPNR rule.

Of course, even the application of one of these rules may not fully define the operation to schedule next. For instance, there may be two operations satisfying the conditions of step 3 and also sharing the same shortest processing time of all such operations. The SPT rule cannot, there-fore, choose between these. Thus it is not sufficient to decide to use just one of these rules to select the scheduled operation at step 3. Rather one must decide upon a hierarchy of rules. Perhaps: first use the MWKR rule; if this does not choose a unique rule, choose amongst those operations selected by the MWKR by the SPT rule; if this does not choose a unique rule, resolve the remaining ambiguity by choosing arbitrarily. However, note first that the RANDOM rule selects a unique operation; and secondly that the RANDOM rule is, of course, equivalent to choosing arbitrarily.

As an example, let us apply each of these priority rules to stage 6 of the active schedule generation given in Table 10.2. There are three operations that might be scheduled at step 3, namely (D,S), (C,S) and (A,S); each requires the *Sun* and each may start before $\phi^* = 10.07$. Here:

the SPT rule selects (A,S), since Algy reads the *Sun* fastest;
the FCFS rule selects (D,S), since (D,S) has been schedulable since stage 0.
the MWKR rule selects (D, S), since Digby has the most reading left;
the LWKR rule selects $(A. S)$, since Algy only has 5 minutes reading left;
the MOPNR rule selects (D,S), since Digby has to read all four papers still;

and

the RANDOM rule selects one of (A,S), (C,S), and (D,S) at random.

Jeremiah, Lalchandani and Schrage (1964) have examined the performance of these and other priority rules empirically. Their results are admirably summarised in Conway, Maxwell and Miller (1967, pp. 121–124) and hence are mentioned briefly here. The authors generated both active and non-delay schedules under each of these priority rules for some 84 different problems. In well over 80% of the cases the non-delay schedules performed better in terms of mean flow time than the active schedules. For minimising maximum flow time the non-delay schedules were again superior, but not as markedly so. The SPT, RANDOM and LWKR rules clearly dominated the rest, for minimising mean flow time, although it must be noted that there were problems in which some other priority rule was superior. For minimising maximum flow time the MWKR rule was a clear winner with SPT a distant second. See also Gere (1966).

Finally in this section on schedule generation, we turn to **probabilistic dispatching or Monte Carlo methods**. So far in this discussion we have considered either enumeration of all the schedules within a class or, at the other extreme, the generation of just one schedule through the choice of a hierarchy of priority rules. In between these extremes we may generate a sample of schedules within the class and then pick the best of these. We might, for instance, use the RANDOM priority rule above, but, instead of generating just one schedule, we might pass through the algorithm, say, 50 times thus producing a sample of schedules. However, in some sense it seems wrong to choose the operation at step 3 simply at random, that is with equal probability from the possible choices. Some choices are both intuitively and empirically better than others. Were this not so, we would not have suggested and discussed the five non-random priority rules above. Yet, if we replace the RANDOM rule with one of these and generate 50 schedules, we shall find that we generate the same schedule 50 times: a rather pointless exercise. What we can do is bias the probabilities used in the random selection to favour those operations that seem the most sensible choice. Thus, returning to stage 6 of our numerical example, we might randomly choose between the operations (D, S), (C, S), (A, S) according to, say, the probabilities 8/12, 3/12 and 1/12 respectively, since we know that empirical evidence favours the MWKR rule for minimising maximum flow-time. Biasing the probabilities like this at each stage means that in generating a sample of schedules we will tend to find those, in some sense, near the unique schedule produced by the deterministic application of the MWKR rule.

We shall follow up this method of probabilistic dispatching no further here save to remark that with a sensible choice of biasing probabilities it does seem to work well, substantially better than the deterministic application of a priority rule. Both Conway, Maxwell and Miller (1967) and Baker (1974) discuss these procedures in much greater depth and report on empirical results.

10.3 MODIFIED OR APPROXIMATE MATHEMATICAL PROGRAMMING METHODS

In Chapter 8 we formulated the general $n/m/P/C_{max}$ as a mixed integer linear programme. There we remarked that this formulation did not really help us towards a solution because integer linear programmes are just as difficult to solve as the original scheduling problem. This difficulty is caused entirely because the formulation constrains some of the variables to be integral. If this constraint was not present, the problem would be easy to solve. For we have the Simplex Method, which solves real linear programmes[†] very quickly. (See e.g. Daellenbach and George, 1979.)

It is interesting to note that the Simplex Method is theoretically an exponential time algorithm. However, its practical performance is excellent; on the majority of practically existing problems it exhibits low order polynomial behaviour. Moreover, Khachian (1979) has recently developed a polynomial time algorithm for linear programmes, but first practical tests indicate that this algorithm has a considerably worse performance for most

[†]i.e. linear programmes in which the variables may take any real value.

problems that the Simplex (*ORSA–TIMS Newsletter*, December 1979). Perhaps this sounds a warning against being too discouraged by NP-completeness results.

But all this is incidental. The point is that we can solve real linear programmes very quickly. So why not solve the linear programming formulation of the scheduling problem without the integral constraints and then take the nearest integral values to the 'optimal' solution found? For general integer programmes this method of solution is capable of producing very poor results (See Garfinkel and Nemhauser (1972)). However, the integer programmes arising from scheduling problems may be so structured that this approach might provide reasonably good solutions. The approach is known as **Rounded Linear Programming**.

Giglio and Wagner (1964) and Ashour (1970a) have investigated this idea empirically. Their results are not particularly encouraging and it should be noted that branch and bound without backtracking (see below) performed better. Ashour found that rounded linear programming produced results which were on average 86% efficient over 100 test examples. Here efficiency is defined as the quotient of the optimal solution and the approximate solution expressed as a percentage; it is the reciprocal of relative performance, which we shall use in Chapter 11. In 5 of the 100 examples efficiencies poorer than 75% were recorded and only in 27 cases was the efficiency better than 95%.

Let us turn now to **branch and bound without backtracking**. A branch and bound search of an elimination tree may take a very long time. But suppose that we do not fully explore the tree. Suppose instead that we use a frontier search and stop as soon as we reach a terminal node representing a complete schedule, i.e. as soon as the trial schedule is set. We might hope that this would find a fairly good schedule. Ashour (1970a, 1970b) has investigated this policy. In fact, he did not use a full frontier search to explore the tree, but rather a variant sometimes called **restricted flooding**. To understand this we refer to Figure 7.9. The search begins by evaluating the lower bounds at all the nodes $\Omega_1, \Omega_2, \Omega_3, \ldots \Omega_\nu$. It then selects the node with the least lower bound, say Ω_i, and evaluates the lower bounds at Ω_{i_1}, $\Omega_{i_2}, \ldots$ Next the lowest lower bound of these, say at Ω_{i_j} is selected and the lower bounds at all the nodes immediately below Ω_{i_j} evaluated. And so on.

Ashour's empirical results for this approach are more promising than those for rounded linear programming. For the same 100 problems he attained an average efficiency of 95%. In no case was the efficiency below 75% and in 76 problems the achieved efficiency was greater than 95%. Nonetheless, there may be practical reasons for preferring rounded linear programming to branch and bound without backtracking. Most computer systems have linear programming packages easily available. Branch and bound algorithms for scheduling problems are less common. Against this it must be remembered that the size of the linear programming problem grows as a quadratic in the number of jobs and may soon exceed the capabilities of all but the largest computers.

10.4 NEIGHBOURHOOD SEARCH TECHNIQUES

An intuitively sensible way to search for a good solution to a scheduling problem is to begin with any feasible schedule, adjust this somewhat, check whether the adjustment has made any improvement, and continue in a cycle of adjustment and testing for improvement until no further progress relative to the performance measure is made. Indeed, we may develop this further. Rather than considering just one adjustment at each stage, we might consider a variety of different adjustments and then pick the most promising of these. This general approach is formalised within a family of heuristic methods known as **neighbourhood search techniques**.

To understand these fully we must consider the concept of a **neighbourhood** of a schedule. To begin with an example, consider a class of problems for which we may concentrate on permutation schedules only, for instance, $n/1//B$ with B regular or $n/3/F/F_{max}$. Also to be specific consider the natural order schedule:

$$(J_1, J_2, J_3, J_4, \ldots, J_{n-1}, J_n)$$

The **single adjacent pairwise interchange neighbourhood** of this is composed of the $(n-1)$ schedules:

$$(J_2, J_1, J_3, J_4, \ldots, J_{n-1}, J_n),$$
$$(J_1, J_3, J_2, J_4, \ldots, J_{n-1}, J_n),$$
$$(J_1, J_2, J_4, J_3, \ldots, J_{n-1}, J_n),$$
$$\ldots \ldots \ldots \ldots \ldots \ldots \ldots \ldots,$$
$$(J_1, J_2, J_3, J_4, \ldots, J_n, J_{n-1}).$$

Thus this neighbourhood is the set of schedules obtained by interchanging the processing order of all adjacent pairs of jobs in the original schedule. Obviously the single adjacent pairwise interchange neighbourhood of any permutation schedule may be defined in a similar fashion. Rather than limit the interchange of jobs to adjacent pairs, we could equally consider the **single pairwise interchange neighbourhood** of a schedule. This consists of the $n(n-1)/2$ schedules obtained by interchanging all pairs of jobs, not just the adjacent pairs. In the general case, where we are not limited to considering just permutations, we define a neighbourhood of a schedule as follows. First we settle upon a mechanism for generating a family of feasible schedules given an initial schedule. In the above examples the mechanisms were respectively adjacent and unlimited pairwise interchange. Given this mechanism—call it X—the **X-neighbourhood** of a schedule is the set of all schedules that may be generated by applying X to this schedule.

Given this concept of an X-neighbourhood, we may write down a general X-neighbourhood search algorithm:

Algorithm 10.3

Step 1 Obtain an initial sequence, called the **seed**.

Step 2 Generate the X-neighbourhood of the seed and check whether any of these schedules improves upon the seed. If none does, stop with the seed as the heuristically selected schedule. Otherwise, continue to *step 3*.

Step 3 Select one of the improved schedules in the X-neighbourhood and replace the seed by this. Return to *step 2*.

Obviously, to apply this algorithm a number of choices have to be made.

(i) How is the initial seed to be chosen?

(ii) What generating mechanism X is to be used?

(iii) How is the replacement seed to be selected in step 3?

The choices (i) and (ii) depend very much upon the class of problem being solved and, by and large, they can only be made as a result of empirical investigation. The choice (iii) depends mainly upon the choice made in (ii). Suppose the generating mechanism produces just a few schedules in the neighbourhood of a seed. In this case it seems sensible to generate them all and select the best in step 3. If, however, the neighbourhood of a schedule is very large, then to even generate them all might take a very long time. In this case it may be best to combine steps 2 and 3. As each schedule is generated, check it for an improvement over the seed. if there is any, immediately select this as the replacement seed. In this fashion the full neighbourhood of a schedule will be generated only rarely, usually only in the very last cycle, in which, of course, there is no improvement.

It is worthwhile emphasising that there is no general reason to expect this neighbourhood search approach to find an optimal solution. Naturally, in some cases it will. Indeed, for certain classes of problem the method with an appropriately chosen generating mechanism is guaranteed to find an optimum. For instance, for the $n/1//\bar{F}$ problem using a random seed and adjacent pairwise interchange neighbourhoods, we must find an optimal schedule. This may be seen from Theorem 3.3, where the optimality of SPT schedules was proved using pairwise interchanges. However, Problem 10.7.8 gives a $3/1//\bar{T}$ problem for which a neighbourhood search fails to find the optimal solution, thus confirming the heuristic status of this approach.

Baker (1974) reports some empirical investigations on the $n/1//\bar{T}$ problem using this approach. He found that by using the EDD sequence as the initial seed and adjacent pairwise interchanges produced very good results, as did using a random seed and all pairwise interchanges. Using a random seed with adjacent pairwise interchanges did not produce good results. Wilkinson and Irwin (1971) give another approach to this problem, which may also be seen as a neighbourhood search technique (Baker, 1974).

Spachis and King (1979) have considered using a local neighbourhood search to solve the general $n/m/G/F_{max}$ problem. Their method ingeniously uses a branch and bound algorithm to enumerate implicitly the neighbourhood of the seed.

10.5 FLOW-SHOP SCHEDULING HEURISTICS

The sections above discuss general heuristic approaches to scheduling. Now we turn to a group of methods designed specifically to tackle flow-shop scheduling and so quite inapplicable to other classes of problem. They are all based upon our physical intuition of the structure of a near optimal schedule in a flow-shop with maximum flow time as the performance measure. Throughout this section we assume that all the ready times are zero.

Remember that in motivating Johnson's algorithm for the $n/2/F/F_{max}$ problem (see especially Fig. 5.2) we argued that the jobs which were placed near the beginning of the schedule should have short processing times on M_1. That way idle time on M_2 would be minimised. Similarly, jobs which were placed near the end of the schedule should have short M_2 processing times in order to minimise M_1 idle time. The methods below carry over this idea to the general $n/m/F/F_{max}$ problem. Namely, it seems intuitively reasonable that jobs placed early in the schedule should have processing times that tend to increase from machine to machine, whereas those towards the end of the schedule should have times that tend to decrease in passing from machine to machine. In this way one may hope that much unnecessary idle time would be avoided.

Before proceeding we should note that all the methods in this section produce permutation schedules. As we have seen if there are 4 or more machines the optimal schedule may not be of this form. We may justify this restriction to permutation schedules on one of two grounds. First, it has been claimed that the optimal permutation schedule does not produce an appreciably worse performance than the optimal general schedule (See Dannenbring, 1977). Second, it is very much easier practically to implement a permutation schedule than a more general one; hence practical problems tend to be of the form $n/m/P/F_{max}$ rather than $n/m/F/F_{max}$.

Palmer (1965) introduced the following heuristic. His idea was to give each job a **slope index** which gives its largest value to those jobs which have the strongest tendency to progress from short to long processing times as they pass from machine to machine. Thus when the jobs are scheduled in decreasing order of slope index we might expect to find a near optimal schedule. For the general $n/m/F/F_{max}$ his slope index for J_i is

$$S_i = -(m-1)p_{i1} - (m-3)p_{i2} - \ldots + (m-3)p_{i(m-1)} + (m-1)p_{im}$$
(10.1)
$$= -\sum_{j=1}^{m} [m - (2j-1)]p_{ij}.$$

To be specific, for an $n/7/F/F_{max}$ (10.1) gives

$$S_i = -6p_{i1} - 4p_{i2} - 2p_{i3} + 0p_{i4} + 2p_{i5} + 4p_{i6} + 6p_{i7}.$$

It may be seen that when the process times tend to increase from machine to machine, the early terms in this sum are small and negative while the later terms are large and positive. Hence this index does have the property suggested.

As an example we apply Palmer's algorithm to the $4/3/F/F_{max}$ problem used in Section 7.3. The table below repeats the data.

Job	**Processing Times**		
J_i	p_{i1}	p_{i2}	p_{i3}
1	1	8	4
2	2	4	5
3	6	2	8
4	3	9	2

Here we find the slope indices:

$$S_1 = -2 \times 1 + 0 \times 8 + 2 \times 4 = \quad 6$$
$$S_2 = -2 \times 2 + 0 \times 4 + 2 \times 5 = \quad 6$$
$$S_3 = -2 \times 6 + 0 \times 2 + 2 \times 8 = \quad 4$$
$$S_4 = -2 \times 3 + 0 \times 9 + 2 \times 2 = -2$$

Thus the algorithm suggests the processing sequence (1, 2, 3, 4) or (2, 1, 3, 4). To decide which to use we would evaluate F_{max} for each of these and choose the better. In fact, both schedules are optimal and share $F_{max} = 28$. (See our calculations in Section 7.3.) Here we are lucky in that optimal solutions are found. Dannenbring (1977) has found empirically for a sample of 1280 small problems ($n \leq 6$, $m \leq 10$). Palmer's heuristic methods finds the optimum in about 30% of the cases.

Campbell, Dudek, and Smith (1970) use a multiple application of Johnson's two machine algorithm to try to obtain a good schedule. Essentially, they create $(m-1)$ scheduling problems; use Johnson's algorithm on each of these; and pick the best of the resulting $(m-1)$ schedules. To be precise, the kth problem ($k = 1, 2, \ldots (m - 1)$) is formed as follows. The constructed processing time on the 1st machine for the ith job is

$$a_i^{(k)} = \sum_{j=1}^{k} p_{ij},$$

that is it is the sum of processing times for the ith job on the first k actual machines. Similarly the constructed time on the second machine is the sum of processing times for the ith job on the last k actual machines:

$$b_i^{(k)} = \sum_{j=m-k+1}^{m} p_{ij}.$$

Note that for $m = 3, k = 2$ this corresponds to an application of Johnson's Algorithm for the special case of the $n/3/F/F_{max}$ problem. Thus Campbell, Dudek and Smith's algorithm may be seen as an intuitive generalisation of this. Note also that this multiple application of Johnson's Algorithm will tend to throw up schedules of the type expected by our intuitive arguments above.

    Applying this algorithm to our example we construct two 2-machine problems corresponding to $k = 1$ and $k = 2$.

Job J_i	Processing Times			Constructed Problem k = 1		Constructed Problem k = 2	
	p_{i1}	p_{i2}	p_{i3}	$a_i^{(1)} = p_{i1}$	$b_i^{(1)} = p_{i3}$	$a_i^{(2)} = p_{i1} + p_{i2}$	$b_i^{(2)} = p_{i2} + p_{i3}$
1	1	8	4	1	4	9	12
2	2	4	5	2	5	6	9
3	6	2	8	6	8	8	10
4	3	9	2	3	2	12	11

Applying Johnson's Algorithm gives the processing sequence (1, 2, 3, 4) for the first constructed problem and (2, 3, 1, 4) for the second. The first we know has $F_{max} = 28$; the second has $F_{max} \geq 29$. So the method would choose the sequence (1, 2, 3, 4) and as Palmer's did, achieve optimality in this case. Dannenbring's empirical results for the 1280 small problems showed that this algorithm achieves optimality roughly 55% of the time.

    Dannenbring (1977) has himself designed a heuristic method, which attempts to combine the advantages of both Palmer's and Campbell, Dudek and Smith's. His idea is to construct just one 2-machine problem on which to apply Johnson's Algorithm, but with the constructed processing times reflecting the same behaviour as Palmer's slope index. The constructed time for the first machine is

$$a_i = mp_{i1} + (m - 1)p_{i2} + \ldots + 1p_{im}$$
$$= \sum_{j=1}^{m} (m - j + 1)p_{ij};$$

and that for the second machine is

$$b_i = p_{i1} + 2p_{i2} + \ldots + mp_{im}$$
$$= \sum_{j=1}^{m} jp_{ij}.$$

Thus a job's constructed time a_i is relatively short if it tends to have short actual processing times on the early machines. Similarly b_i is relatively short if J_i tends to have short actual processing times on the later machines.

Applying this method to our example:

	Processing times			Constructed Processing Times	
J_i	p_{i1}	p_{i2}	p_{i3}	$a_i = 3p_{i1} + 2p_{i2} + p_{i3}$	$b_i = p_{i1} + 2p_{i2} + 3p_{i3}$
1	1	8	4	23	29
2	2	4	5	19	25
3	6	2	8	30	34
4	3	9	2	29	27

Thus an application of Johnson's Algorithm gives the processing sequence $(2, 1, 3, 4)$ which has $F_{max} = 28$ and again is optimal. Dannenbring's empirical results show that for small problems this method finds the optimal schedule in about 35% of all cases.

Dannenbring compared these three algorithms and some others not discussed here on many problems, both large and small. Overall he found that Campbell, Dudek and Smith's Algorithm performed much better than the other two. The most important result of Dannenbring's study is that he found that all these methods could be dramatically improved by using their output as initial seeds in a neighbourhood search routine, even if one limited the search to just one cycle. This is perhaps the most promising direction to investigate in the design of flow shop heuristics.

It is worth emphasising the purpose of our study in this section. Undoubtedly these three algorithms are important contributions to the heuristic solution of flow-shop problems. However, the solution of such problems was not our direct concern. Indeed, had it been, we should have also discussed the contributions of, among others, Ashour (1967, 1970a),

Gupta (1971, 1972), and King and Spachis (1980). Rather our intention has been to illustrate heuristic methods that derive from our understanding of particular problems and our consequent intuition of the structure of good schedules.

10.6 REFERENCES AND FURTHER READING

Conway, Maxwell and Miller (1967), Baker (1974), and Coffman (1976) all discuss various heuristic methods that may be used in scheduling. Silver *et al.* (1980) discuss and survey general approaches to heuristic problem-solving in all areas of operational research.

In Section 10.2 we briefly mentioned probabilistic dispatching methods, in which schedules are generated randomly, perhaps with a sensible bias, and the best selected. This approach is part of a general heuristic method of combinatorial optimisation known as Monte Carlo. In any method which generates a random sample and selects the best there are two natural and related questions: how large should the sample be and how near optimal is the best schedule found likely to be? Randolph *et al.* (1973) have addressed this question within the context of scheduling, and Rinnooy Kan (1976) has extended their approach.

The algorithms of Section 10.5 are all essentially sorting methods; they assign an index to each job and then sort the jobs into the order given by these indices. (Although Johnson's Algorithm, and hence the approach of Campbell *et al.*, does not apparently take this form, it can be shown to be equivalent to sorting jobs according to the index sign $(a_i - b_i)/\min\{a_i, b_i\}$. See Problem 10.7.11.) Page (1961) has discussed the relationship between such approaches and general sorting algorithms.

Holloway and Nelson (1973, 1974a, 1975) in an interesting series of papers have developed a heuristic approach to a general job-shop problem formulated with slightly different assumptions from those we have adopted.

10.7 PROBLEMS

1. Show that any optimal schedule for an $n/m/G/B$ problem where the measure of performance B is regular can be modified to an active schedule without increasing the value of B.

2. Show that a non-delay schedule is necessarily an active schedule.

3. Show that the possible choice of M^* in step 2 of Algorithm 10.1 may be made arbitrarily.

4. Show that any given active schedule may be generated by an appropriate sequence of choices at step 3 of Algorithm 10.1.

5. Show that Algorithm 10.2 does generate a non-delay schedule and

that any given non-delay schedule may be generated by an appropriate sequence of choices at step 3.

6. Generate a non-delay schedule for the newspaper reading problem using a hierarchy of priority rules of your choice. Compare your result with the unique optimal solution.

7. Using branch and bound without backtracking based upon a restricted flooding search, find an approximate solution to the $6/3/F/F_{max}$ example posed in Problem 7.6.8.

8. Consider the $3/1//\bar{T}$ problem with data

Job	Processing Time	Due Date
1	1	4
2	2	2
3	3	3

Begin with the seed (3,1,2) and use the neighbourhood search technique based upon adjacent pairwise interchange. Confirm by evaluating $\bar{T}$ for all 6 possible schedules that you have not found the optimal schedule.

9. Beginning with the seed (1, 2, 3, 4) use a neighbourhood search technique based upon all pairwise interchanges on the $n/1//\bar{T}$ problem used as an example in Section 6.2.

10. Apply each of the three heuristic methods discussed in Section 10.5 to the flow-shop examples in Problems 7.6.2 and 7.6.3.

11. Show that Johnson's Algorithm is equivalent to sorting jobs into increasing order of the index

$$\frac{\text{sign}(a_i - b_i)}{\min\{a_i, b_i\}} ,$$

where

$$\text{sign}(a_i - b_i) = \begin{cases} -1 & \text{if } (a_i - b_i) < 0, \\ +1 & \text{if } (a_i - b_i) \geq 0. \end{cases}$$

Chapter **11**

Heuristic Methods: Worst Case Bounds

11.1 INTRODUCTION

In the previous chapter we introduced the idea of heuristic methods. We referred to some empirical studies which showed that in the majority of cases these 'sensible' rules generate reasonably good schedules. However, they cannot guarantee optimality and there is always a nagging doubt that in unfortunate cases they may lead to very poor results. Our first objective in this chapter is to consider worst case bounds on the performance of certain heuristic algorithms, i.e. to discover how poorly they may perform in the worst of possible circumstances. We shall not investigate many heuristic methods, but rather restrict ourselves to a few that exhibit certain typical patterns of worst case behaviour.

We are driven to use heuristic methods because many scheduling problems are NP-hard; so there are unlikely to be polynomial time algorithms for finding optimal schedules. Reluctantly, therefore, we have stopped seeking the best possible solution and agreed to accept one that is not too poor. Once that we have introduced the concept of a worst case bound on the sub-optimality of heuristically generated schedules, it is only natural that we should ask how well can we do in polynomial time; it does not make sense to avoid an exponential time optimisation algorithm by using an approximate method that takes as long. Given that we limit ourselves to polynomial time heuristic algorithms, what sort of performance can we guarantee? We shall find that the theory of NP-completeness again presents us with a very pessimistic outlook.

Assumptions for Chapter 11
Our main assumption will be that all the numerical quantities which define an instance are integers.

It should also be noted that the class of scheduling problems considered in the next section breaks many of the assumptions of Chapter 1.

11.2 SCHEDULING INDEPENDENT JOBS ON IDENTICAL MACHINES

For this section there is some advantage to be gained by considering an entirely distinct class of scheduling problems from that we have discussed before. Doing so has the advantage that for this one class we can find heuristic algorithms that exhibit each of the different possible patterns of worst case behaviour that we wish to discuss. Thus we shall consider the problem of scheduling n independent jobs $\{J_1, J_2, \ldots, J_n\}$ on m identical machines $\{M_1, M_2, \ldots, M_m\}$. Each job J_i requires only one processing operation and that operation may occur on any of the m machines. Since the machines are identical, the processing time p_i of job J_i is the same whichever machine is used. The jobs may be processed in any order and are available at time zero (all $r_i = 0$). No machine may process more than one job at a time. Our objective is simply to minimise C_{max}, the time that the last job is completed.

As an example of this problem, consider a typing pool with m secretaries all with identical typing speeds. At the beginning of a day the supervisor's in-tray contains n pieces of work and her task is to assign this to the typists so that it is completed as soon as possible. Of course, in real life the typists do have different speeds (non-identical machines), some jobs have to be done before others (precedence constraints), and the jobs do not all arrive together (non-zero ready times or stochastic problems). Following our usual practice, we shall ignore such complications. For $m > 1$ this problem is NP-complete; so it does make sense to consider heuristic methods. (When $m = 1$ this problem becomes the trivial $n/1//C_{max}$ problem, for which C_{max} is a constant independent of the schedule.)

We shall need some more notation. Recall that an instance, I, of the problem is a specification of the data $n, m, p_1, p_2, \ldots, p_n$. We define OPT($I$) to be the maximum completion time of all the jobs if an optimal schedule is used for instance I. $A(I)$ is the maximum completion time if a particular heuristic algorithm A is used for instance I. We shall look for bounds on the **relative performance**, $R_A(I) = A(I)/\text{OPT}(I)$. Note that $R_A(I) \geq 1$ for all A and for all I. Moreover, the better the schedule found by A the nearer $R_A(I)$ is to 1.

The first heuristic method that we shall consider is the **List Scheduling Algorithm**. This lists the jobs in any order and then assigns them in this order to the machines as they become free. Note that the list of jobs is in a completely arbitrary order. The idea can most easily be assimilated through an example.

For six jobs, three machines and the processing times:

$$p_1 = 8, p_2 = 6, p_3 = 2, p_4 = 8, p_5 = 1, p_6 = 4,$$

suppose that we use the list $(J_4, J_2, J_5, J_3, J_1, J_6)$. At $t = 0$ all machines are available for processing so we assign J_4 to M_1, J_2 to M_2, and J_5 to M_3. Since more than one machine was available, we have made the conventional arbitrary choice of assigning jobs to machines in order of increasing machine index. The first machine to finish its task is M_3 at $t = 1$, so we assign J_3 to this next. Again M_3 finishes before the others, so we assign J_1 to it. Next at $t = 6$, M_2 completes and is assigned J_6. All processing eventually completes at $t = 11$. This is illustrated in the Gantt diagram in Fig. 11.1. Our first result shows that this algorithm guarantees that $R_A(I) \leq 2 - 1/m$ for all instances I.

Theorem 11.1. (Graham, 1969) If A is the list scheduling algorithm above, then

$$R_A(I) \leq 2 - 1/m \quad \text{for all instances } I.$$

Proof First we find two lower bounds on OPT(I). Clearly

$$\text{OPT}(I) \geq \max_{i=1}^{n}\{p_i\}. \tag{11.1}$$

Also a little thought shows that in the best possible circumstances the processing is divided equally between the machines. So

$$\text{OPT}(I) \geq \frac{1}{m} \sum_{i=1}^{n} p_i. \tag{11.2}$$

Let H_j be the time that processing finally halts on machine M_j. Let J_k be a job that completes at $A(I)$, that is J_k completes last. Since the list scheduling algorithm always assigns the next job to a machine as soon as it finishes processing its current job, no machine can cease processing while there are

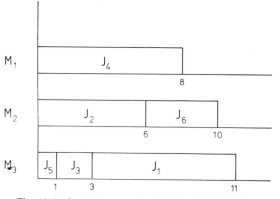

Fig. 11.1 Gantt diagram for the list scheduling example.

any unassigned jobs in the list. Thus none can cease before the processing of J_k starts:

$$H_j \geq A(I) - p_k \quad \text{for all } M_j.$$

Moreover, we know that at least one machine halts at $A(I)$. So

$$\sum_{i=1}^{n} p_i = \sum_{j=1}^{m} H_j \geq (m-1)(A(I) - p_k) + A(I).$$

Hence $$A(I) \leq \frac{1}{m} \sum_{i=1}^{n} p_i + \frac{(m-1)}{m} p_k$$

$$\leq \text{OPT}(I) + \frac{(m-1)}{m} \text{OPT}(I), \quad \text{by (11.1) and (11.2)}.$$

Therefore $$R_A(I) = \frac{A(I)}{\text{OPT}(I)} \leq 2 - \frac{1}{m} \quad \text{as required}.$$

In fact we can show that this bound is tight, that is there exists an instance and a list such that $R_A(I) = 2 - 1/m$. Consider an instance with m machines and $(2m - 1)$ jobs. Let the processing times be

$$p_i = \begin{cases} (m-1) & i = 1, 2, \ldots, (m-1), \\ 1 & i = m, (m+1), \ldots, (2m-2), \\ m & i = (2m-1). \end{cases}$$

If we use the natural order list $(J_1, J_2, \ldots, J_{(2m-1)})$ in the list scheduling algorithm, we obtain the schedule shown in Fig. 11.2(a). This has $A(I) = (2m - 1)$. If on the other hand we use the list $(J_{(2m-1)}, J_1, J_2, \ldots, J_{(2m-2)})$, then we obtain the schedule shown in Fig. 11.2(b). This is clearly optimal, since the processing is divided evenly between the machines. Thus $\text{OPT}(I) = m$. Hence here $R_A(I) = A(I)/\text{OPT}(I) = (2m-1)/m = 2 - 1/m$, and we have an instance and a list which attains the bound in Theorem 11.1. For obvious reasons we call the bound provided by Theorem 11.1 a **performance guarantee** and we call this style of analysis **worst case analysis**.

As an illustration we apply this theorem to the example of a typing pool mentioned above. Suppose that there are four typists with identical speeds and that the supervisor assigns work to them by the list scheduling algorithm with an arbitrarily ordered list, say the order of the work in the in-tray. Then the overall time to complete the work could be up to $2 - \frac{1}{4} = 1\frac{3}{4}$, as long as it would be if an optimal assignment were used. In other words, Theorem 11.1 provides a guarantee that in the worst possible case assigning work in the order of the in-tray can lead to a delay in completion relative to the optimal of no more than 75%.

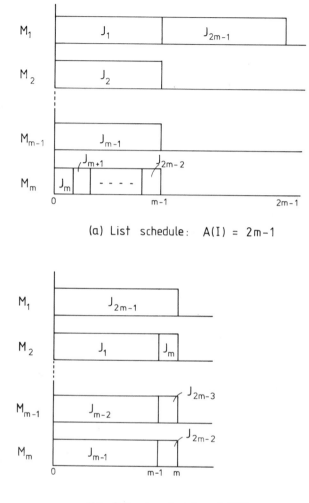

(a) List schedule: A(I) = 2m-1

(b) Optimal schedule: Opt(I) = m

Fig. 11.2 An example showing that the bound in Theorem 11.1 is the best possible.

An obvious question is whether there are other algorithms which give better performance guarantees than the list scheduling algorithm; and, in fact, there are. A natural improvement is to order the jobs in the list in more sensible fashion. A glance at Fig. 11.2 suggests that the poor performance of the algorithm in the first case is due to the longest job occurring last in the list. Thus it is natural to consider the **longest processing time algorithm**. This simply modifies the list scheduling algorithm by

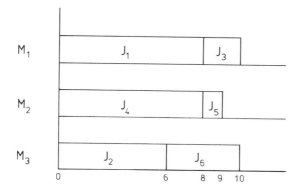

Fig. 11.3 Gantt diagram for the schedule generated by the longest processing time algorithm.

demanding that the jobs are ordered in non-increasing processing time in the list. For instance, if we consider the 6 job machine example above, we see that a non-increasing list of jobs is $(J_1, J_4, J_2, J_6, J_3, J_5)$. This gives the schedule show in Fig. 11.3. Because all the p_i are integral and because $\frac{1}{3}\sum_{i=1}^{6}p_i = 9\frac{2}{3}$, this schedule must be optimal. However, the longest processing time schedule does not always find the optimal schedule. Graham (1969) has shown by means of a long, involved proof, which we shall not present, that the following performance is guaranteed.

Theorem 11.2 (Graham, 1969). If A is the longest processing time algorithm, then

$$R_A(I) \leq \frac{4}{3} - \frac{1}{3m} \quad \text{for all instances } I.$$

If we return to the typing pool example, we may see the improvement that this algorithm brings. Suppose that the supervisor now assigns work to the typists in order of non-increasing processing time; then the work will be completed within $(4/3 - 1/12) = 1\frac{1}{4}$ times that resulting from the optimal assignment.

The above development illustrates one stage in a cyclical process by which worst case analysis can lead to better and better heuristic algorithms. Initially an apparently sensible algorithm is proposed, here the list scheduling algorithm. Guarantees are found upon its performance (Theorem 11.1) and, more importantly for algorithm design and improvement, examples which exhibit its weaknesses are sought (Fig. 11.2). Improvements are made to the algorithm to counter these; here we developed the longest processing time algorithm. Performance guarantees are sought upon this new algorithm to ensure that, in the sense of the bounds, there has indeed been an improvement. Given that there has, the new algorithm becomes a candidate for improvement and the process cycles onward.

This process of algorithm design and improvement is somewhat piecemeal and we do not know that we can continue it to produce a heuristic method that guarantees a particular desired result. However, for some problems it is possible to refine heuristic algorithms in a far more methodical fashion. In particular, given a desired accuracy, ε, it is possible to find an algorithm A that guarantees $R_A(I) \leq 1 + \varepsilon$ for all I.

In the following we fix the number of machines as $m = 2$. Consider the following algorithm A_k for any chosen integer $k > 0$.

Step 1 Select the k jobs with the longest processing times; if $k > n$, select all n jobs.

Step 2 Ignoring the unselected jobs, schedule these k jobs optimally using complete enumeration.

Step 3 Append the remaining $(n - k)$ jobs to the schedule by using the list scheduling algorithm with these jobs arbitrarily ordered.

Theorem 11.3 (Graham, 1969). If A_k is the algorithm above, then

$$R_{A_k}(I) \leq 1 + \frac{\frac{1}{2}}{[1 + (k/2)]} \text{ for all instances } I,$$

where $[x]$ denotes the greatest integer not exceeding x.

Proof. See Problem 11.6.2.
Clearly given any $\varepsilon > 0$ we can find an integer k such that

$$\frac{\frac{1}{2}}{[1 + (k/2)]} < \varepsilon.$$

Hence we can design an algorithm to achieve any desired accuracy.

However, before we are blinded by our success, let us consider the computation required in applying A_k to an instance with n jobs. Selecting the k longest jobs takes at most $O(n^2)$ operations: see Section 9.2 where we counted the number of operations necessary to apply the SPT rule. Finding the optimal schedule for the k selected jobs by complete enumeration takes $O(2^k)$ operations; for there are 2^k possible schedules, since each job may be assigned to either the first machine or the second. List scheduling the remaining $(n - k)$ jobs takes $O(n - k)$ operations. So in total the algorithm requires $O(n^2) + O(2^k) + O(n - k) = O(n^2 + 2^k)$ operations. Since the number of jobs cannot exceed the instance size, the time complexity of A_k is at most $O(v^2 + 2^k)$. (Our analysis has been rather primitive, although quite sufficient for our illustrative purposes. The number of operations required by A_k may be reduced slightly by various techniques; see Aho *et al.*, 1974.)

Thus we have a sequence of algorithms $A_1, A_2, \ldots, A_k, \ldots$ with polynomial time complexity for each fixed k. However, the constant term in the polynomial is $O(2^k)$ and this means that for except for small k the algorithms will be of little practical use (See Table 9.1). Nonetheless, we do have here a scheme for generating heuristic algorithms of any desired accuracy and, once this accuracy has been specified, we know that the algorithm will run in polynomial time. In theoretical terms this is a major step forward. We call such a sequence of algorithms a **polynomial time approximation scheme**, since all the algorithms take the same form and each does run in polynomial time.

The difficulty is, of course, that, although the time complexity is polynomial in the problem size, it is exponential in the desired accuracy $1/\varepsilon$, because k is $O(1/\varepsilon)$. What we should really like is an approximation scheme that produces algorithms that have time complexity that is polynomial both in the problem size and the desired accuracy. Specifically we shall say that a sequence of algorithms $A_1, A_2, \ldots$ form a **fully polynomial time approximation scheme** if the time complexity function of A_k is polynomial both in the problem size and in $1/\varepsilon$, where the performance guarantee of A_k is

$$R_{A_k}(I) \leqslant 1 + \varepsilon \quad \text{for all instances } I.$$

It is surprising to find that for certain scheduling problems, including the one here, such approximation schemes exist.

To develop a fully polynomial approximation scheme for scheduling independent jobs on two identical machines, we must first consider a dynamic programming method, which solves this problem optimally.

Suppose that we have n jobs with processing times $p_1, p_2, \ldots, p_n$ to be processed on either of two identical machines. Let $b = \Sigma_{i=1}^n p_i$. We know that $\text{OPT}(I) \geqslant b/2$ for any instance. Our aim will be to find a subset S of $\{J_1, J_2, \ldots, J_n\}$ such that

(i) $\displaystyle\sum_{J_i \text{ in } S} p_i \leq \frac{b}{2} ;$ and

(11.3)

(ii) if S' is a subset of $\{J_1, \ldots, J_n\}$ such that $\Sigma_{J_i \text{ in } S'} p_i \leq b/2$,

then $\displaystyle\sum_{J_i \text{ in } S'} p_i \leq \sum_{J_i \text{ in } S} p_i.$

Thus S is a subset of jobs whose total processing time is less than half the total processing time of all the jobs and which has the largest total time of all such subsets. Once we have found S, we may obtain an optimal schedule by assigning the jobs in S to one machine and the remaining jobs to the other. The maximum completion time of this schedule is $(b - \Sigma_{J_i \text{in} S} p_i)$ and this exceeds $b/2$ by the least amount possible.

Define $T(i, j)$ for integers i and j with $1 \leqslant i \leqslant n$ and $1 \leqslant j \leqslant b/2$ by

$$T(i, j) = \begin{cases} 0 & \text{if there does not exist a subset } U \text{ of } \{J_1, J_2, \ldots, J_i\} \text{ such that} \\ & \sum_{J_l \text{ in } U} p_l = j; \\ k & \text{if such a subset } U \text{ exists and } k \text{ is the maximum subscript} \\ & \text{such that } J_k \text{ is in } U. \end{cases}$$

Thus $T(i, j)$ indicates whether there is a subset of the first i jobs such that their total processing time sums to j. Moreover, if there is such a subset U, $T(i, j)$ indexes the last job in U. Although there may be more than one such subset U we shall see that $T(i, j)$ is defined in a natural and unambiguous manner.

Clearly $T(1, j)$ is given by

$$T(1, j) = \begin{cases} 0 & \text{if } j \neq p_1, \\ 1 & \text{if } j = p_1. \end{cases} \tag{11.4}$$

We may develop a general recursion for $T(i + 1, j)$ very simply. Either there is a suitable subset U with J_{i+1} last or, if a subset U exists, it is based upon the first i jobs alone. Thus

$$T(i + 1, j) = \begin{cases} i + 1 & \text{if either } j = p_{i+1} \text{ or } T(i, j - p_{i+1}) \neq 0 \\ T(i, j) & \text{otherwise} \end{cases} \tag{11.5}$$

Hence we may generate all $T(i, j)$ by using (11.4) first to find $T(1, j)$ for all j, and then using (11.5) recursively to find $T(2, j)$ for all j, then $T(3, j)$ for all j, and so on. Once all the $T(i, j)$ have been generated we simply find the largest j such that $T(n, j) \neq 0$. This gives $\Sigma_{J_i \text{in} S} p_l$ for a subset S which obeys conditions (i) and (ii) of (11.3). We can find the jobs that lie in S easily from the $T(i, j)$. (See Problem 8.6.3.) The optimal schedule is obtained, as indicated above, by assigning the jobs in S to one machine and the remaining jobs to the other.

The number of operations required by this dynamic programming algorithm is proportional to the number of $T(i, j)$. There are $n[b/2]$ of these so the algorithm has time complexity $O(nb)$. Despite appearances, this does not mean that it is a polynomial time algorithm. Remember that we have insisted that any reasonable encoding convention does not represent numbers to the base 1. Hence the problem size increases logarithmically with the magnitude of the data. Conversely the magnitude of the data, and hence $b = \Sigma_{i=1}^{n} p_i$, can increase exponentially with the problem size. Thus the time complexity function of this algorithm being $O(nb)$ indicates exponential behaviour. However, whether it is a polynomial time algorithm or not is not strictly our concern here. Our intention is to develop a fully polynomial time approximation scheme and to do that we shall simply need the knowledge that these problems can be solved optimally in $O(nb)$ operations.

For each integer $k > 0$ define a heuristic algorithm A_k as follows. Replace all the processing times that define I by

$$\tilde{p}_i = [p_i/k]$$

where, as usual, $[x]$ is the greatest integer not exceeding x. Solve this derived problem using the dynamic programming algorithm above. Use the resulting schedule for the original problem.

Now let us obtain a performance guarantee for this algorithm. We shall first show that

$$A_k(I) - \text{OPT}(I) \leqslant nk.$$

Let S_0 be the subset of jobs assigned to the first machine under the optimal schedule for the *original* problem. Let S_k be the subset assigned to the first machine under the optimal schedule for the *modified* problem, that is S_k are those jobs assigned by A_k. Then we have:

$$\text{OPT}(I) = \max\left\{ \sum_{J_i \text{ in } S_0} p_i, \sum_{J_i \text{ not in } S_0} p_i \right\}$$

and

$$A_k(I) = \max\left\{ \sum_{J_i \text{ in } S_k} p_i, \sum_{J_i \text{ not in } S_k} p_i \right\}.$$

Next we let

$$X = \max\left\{ \sum_{J_i \text{ in } S_k} \tilde{p}_i, \sum_{J_i \text{ not in } S_k} \tilde{p}_i \right\}$$

and

$$Y = \max\left\{ \sum_{J_i \text{ in } S_0} \tilde{p}_i, \sum_{J_i \text{ not in } S_0} \tilde{p}_i \right\};$$

in other words, X and Y are the maximum completion times under S_k and S_0 respectively in the modified problem. Remember that A_k constructs S_k to be optimal for this problem. Hence

$$X \leqslant Y. \tag{11.6}$$

Since $0 \leqslant (p_i/k) - \tilde{p}_i < 1$, we have $0 \leqslant (p_i - k\tilde{p}_i) < k$. Hence, because there are at most n jobs in any subset,

$$0 \leq \sum_{J_i \text{ in } S} (p_i - k\tilde{p}_i) < nk \quad \text{for any subset } S. \tag{11.7}$$

Consider $(A_k(I) - kX)$:

$$(A_k(I) - kX) = \max\left\{ \sum_{J_i \text{ in } S_k} p_i, \sum_{J_i \text{ not in } S_k} p_i \right\} - kX$$

$$= \max\left\{ \left(\sum_{J_i \text{ in } S_k} p_i \right) - kX, \left(\sum_{J_i \text{ not in } S_k} p_i \right) - kX \right\}.$$

But from the definition of X we have both $X \geq \sum_{J_i \text{ in } S_k} \tilde{p}_i$ and $X \geq \sum_{J_i \text{ not in } S_k} \tilde{p}_i$.

Hence $\quad (A_k(I) - kX) \leq \max\left\{ \sum_{J_i \text{ in } S_k} (p_i - k\tilde{p}_i), \sum_{J_i \text{ not in } S_k} (p_i - k\tilde{p}_i) \right\}$

$$\leq nk \quad \text{by (11.7).}$$

Also consider $(\text{OPT}(I) - kY)$:

$$(\text{OPT}(I) - kY) = \text{OPT}(I) - k \max\left\{ \sum_{J_i \text{ in } S_0} \tilde{p}_i, \sum_{J_i \text{ not in } S_0} \tilde{p}_i \right\}$$

$$= \min\left\{ \text{OPT}(I) - \sum_{J_i \text{ in } S_0} k\tilde{p}_i, \text{OPT}(I) - \sum_{J_i \text{ not in } S_0} k\tilde{p}_i \right\}.$$

But from the definition of $\text{OPT}(I)$ we have both $\text{OPT}(I) \geq \sum_{J_i \text{ in } S_0} p_i$ and $\text{OPT}(I) \geq \sum_{J_i \text{ not in } S_0} p_i$.

Hence $\quad (\text{OPT}(I) - kY) \geq \min\left\{ \sum_{J_i \text{ in } S_0} (p_i - k\tilde{p}_i), \sum_{J_i \text{ not in } S_0} (p_i - k\tilde{p}_i) \right\}$

$$\geq 0 \quad \text{by (11.7).}$$

Hence $\quad (A_k(I) - kX) - (\text{OPT}(I) - kY) \leq nk - 0.$

$$A_k(I) - \text{OPT}(I) \leq nk - k(Y - X).$$

Thus, as required,

$$A_k(I) - \text{OPT}(I) \leq nk \quad \text{by (11.6).}$$

Now that we have this bound on the difference between $A_k(I)$ and $\text{OPT}(I)$, it is a relatively simple matter to develop a performance guarantee. Dividing through by $\text{OPT}(I)$ gives

$$\frac{A_k(I)}{\text{OPT}(I)} \leq 1 + \frac{nk}{\text{OPT}(I)}.$$

But we know that $\text{OPT}(I) \geq \frac{1}{2} \sum\limits_{i=1}^{n} p_i$, so

$$R_{A_k}(I) = \frac{A_k(I)}{\text{OPT}(I)}$$

$$\leq 1 + \frac{2nk}{\sum\limits_{i=1}^{n} p_i}$$

This method of finding approximate solutions is known as **rounded dynamic programming** since it finds a schedule by solving a derived problem with processing times rounded to multiples of k. Here we have proved:

Theorem 11.4. (Sahni, 1976). If A_k is the rounded dynamic programing heuristic method discussed above then

$$R_{A(k)} \leq 1 + \frac{2nk}{\sum\limits_{i=1}^{n} p_i} \quad \text{for all instances } I.$$

Suppose that we wish to achieve $R_{A_k}(I) \leq 1 + \varepsilon$. Clearly putting $k = \left\lceil \dfrac{\varepsilon \sum\limits_{i=1}^{n} p_i}{2n} \right\rceil$ will achieve this. Moreover, the time complexity of A_K is

$$O\left(n \sum_{i=1}^{n} \tilde{p}_i \right) = O\left(\frac{n}{k} \sum_{i=1}^{n} p_i \right)$$

$$= O\left(\frac{n^2}{\varepsilon} \right)$$

which is polynomial in n and $1/\varepsilon$. So we have found a fully polynomial time approximation scheme for our problem.

11.3 PERFORMANCE GUARANTEES AND NP-COMPLETENESS

In the last section we introduced the concept of performance guarantees for heuristic schedules. Having done so, we began to consider how much computation was necessary to guarantee a desired accuracy. Ideally we should like to find a fully polynomial time approximation scheme for each NP-hard scheduling problem; because then, even though we cannot solve the problem optimally in polynomial time, we can at least approximate it without the computational requirements increasing exponentially with either the problem size or the desired accuracy. In the following we shall discover that the theory of NP-completeness does not encourage us in our hope. Unless $P = NP$, there can be no fully polynomial time approximation schemes for many scheduling problems. Indeed, for some problems there is a constant $\Delta > 0$ such that we cannot achieve a relative performance $R_A(I) \leq 1 + \varepsilon$ for any $0 \leq \varepsilon < \Delta$ without using an exponential time algorithm.

Let us begin with a simple theorem. We have always considered guarantees for the relative performance

$$R_A(I) = \frac{A(I)}{\mathrm{OPT}(I)}$$

Would it be too hopeful to search for polynomial time heuristic algorithms which guarantee a bound on the absolute error? Yes, it would

Theorem 11.5 (Garey and Johnson, 1979). If $P \neq NP$, no polynomial time algorithm A for an NP-hard scheduling problem can guarantee that for some fixed K

$$|A(I) - \mathrm{OPT}(I)| \leq K \quad \text{for all instances } I.$$

Proof. Remember that we are assuming that all numerical quantities in the definition of the instance are integral. Suppose that such a $K > 0$ and such an algorithm exists. We may assume K is integral. From an instance I create a new instance I' by multiplying all processing times, due dates, ready times etc. by $(K + 1)$, *but* do not change the number of jobs or the number of machines. Clearly the optimal schedules for I and I' are identical, and we have

$$(K + 1)\mathrm{OPT}(I) = \mathrm{OPT}(I').$$

Note that all quantities in I' are integer multiples of $(K + 1)$. Hence

$$|A(I') - \mathrm{OPT}(I')| \leq K \Rightarrow A(I') = \mathrm{OPT}(I').$$

Thus A solves I' optimally in polynomial time and so we can solve I optimally in polynomial time. Therefore, either the problem is not NP-hard or $P = NP$, and the theorem is proved.

So it was not accidental that we only considered guarantees on relative performance. We could not usefully have done otherwise.

Next we rule out the existence of fully polynomial time approximation schemes for very many scheduling problems. It will help you if you look back to the definition of NP-hardness in the strong sense given at the end of Section 9.3.

Theorem 11.6 (Garey and Johnson, 1978). Let v_I be the size of a scheduling instance I and η_I be an upper bound on the magnitudes of the data. Suppose there exists a polynomial $q(v, \eta)$ such that

$$\text{OPT}(I) < q(v_I, \eta_I) \quad \text{for all instances } I.$$

Then the existence of a fully polynomial time approximation scheme implies the existence of a pseudo-polynomial time optimisation algorithm.

Proof. Suppose $A_1 A_2 A_3 \ldots$ is such an approximation scheme. Thus for any $\varepsilon > 0$ we can find k_ε such that A_{k_ε} guarantees

$$\frac{A_{k_\varepsilon}(I)}{\text{OPT}(I)} \leq 1 + \varepsilon, \tag{11.8}$$

and also that A_{k_ε} runs in $O(p(v_I, 1/\varepsilon))$ time for some polynomial p. For a given instance we set $\varepsilon = 1/q(v_I, \eta_I)$ and choose the appropriate A_{k_ε}. Hence we can find the approximate solution in

$$O(p(v_I, 1/\varepsilon)) = O(p(v_I, q(v_I, \eta_I))),$$

i.e. in pseudo-polynomial time.
   From (11.8) we have:

$$A_{k_\varepsilon}(I) - \text{OPT}(I) \leq \text{OPT}(I)\varepsilon$$

$$= \frac{\text{OPT}(I)}{q(v_I, \eta_I)}$$

$$< 1, \text{ since } \text{OPT}(I) < q(v_I, \eta_I) \text{ by assumption.}$$

Since all numbers are integers, $A_k(I) = \text{OPT}(I)$. So we have a pseudo-polynomial time optimisation algorithm.

Corollary 11.7. If the scheduling problem in Theorem 11.6 is NP-hard in the strong sense, then no fully polynomial time approximation scheme can exist for it, unless P = NP.

Proof. This corollary follows trivially from the definition of NP-hard in the strong sense.
   The condition that $\text{OPT}(I) < q(v_I, \eta_I)$ is trivially obeyed for very many scheduling problems. For instance, $C_{\max} < v_I \eta_I$ for any processing sequence since the number of data is less than v_I and all the processing times and ready times are less than η_I. Thus the deciding factor for the existence of a fully polynomial approximation scheme is generally whether the problem is NP-hard in the strong sense or not. It will come as no surprise to find that the job-shop and the flow-shop problem with $m \geq 3$ are both NP-hard in the strong sense. So too are the $n/1//\bar{C}$ problem with the non-zero ready times and the $n/1//\Sigma \gamma_i$ problem, whether or not the ready times are zero.
   Finally, we consider a theorem which for some problems put a lower bound upon the relative performance achievable in polynomial time.

Theorem 11.8 (Garey and Johnson, 1979). For some scheduling problem and some fixed $K > 0$ independent of the instance, suppose that the recognition problem

'given I, is $\text{OPT}(I) \leq K$?'

is NP-complete. Then, if $P \neq NP$, no polynomial time approximation algorithm can satisfy

$$\frac{A(I)}{\text{OPT}(I)} < 1 + \frac{1}{K}$$

Proof. See Problem 11.6.6.

If precedence constraints are added to the problem of scheduling independent jobs on identical machines, then for $K = 3$ the recognition problem in Theorem 11.8 is NP-complete. So we cannot find any polynomial time heuristic algorithm A that guarantees $R_A(I) \leq 4/3$.

11.4 WORST CASE ANALYSIS AND EXPECTED PERFORMANCE

Again the theory of NP-completeness has provided us with a very pessimistic outlook. Already we have learned that for many scheduling problems there is little chance of developing polynomial time optimising algorithms. Now we discover that there is equally little chance of developing satisfactory approximation schemes. However fascinating the theory of NP-completeness may be, few can study it without becoming dejected.

But let us try to be more positive. As we have already remarked, practical scheduling problems cannot be left unsolved; some processing sequence must be used and, depending on that sequence, costs will be incurred. Surely we can do better than simply pick a schedule at random. So suppose that we are faced with a large NP-hard problem. What should we do?

First, we should remember that the theory of NP-completeness applies to classes of problems. We are faced with a particular instance. Branch and bound or some other form of implicit enumeration may solve this instance quickly. (Remember than the exponential time Simplex method is almost always quicker than Khachian's polynomial time algorithm.) Thus despite all our fears we may find an optimal schedule. Nonetheless, for the majority of NP-hard problems we shall fail. Thus we shall be driven to try heuristic methods.

Here again we should remember that our theory applies to classes of problems and that we are faced with a particular instance. Worst case bounds are, by definition, only attained in the worst of possible circumstances. In our particular instance a heuristic algorithm with a poor worst case bound may, in practice, perform very well. Indeed it is known empirically that the average performance of some heuristic methods is far better than their worst case analysis might suggest (Garey and Johnson, 1979; Sahni, 1977). For example, in one study it was found that worst case bounds on $R_A(I)$ of 1.70 corresponded to an average value of 1.07. In this respect it should be noted that Karp (1975) has begun the very difficult task of producing theoretical, rather than empirical, descriptions of the expected behaviour of certain algorithms. Thus we should try whatever heuristic algorithms are available. Moreover, we should examine our problem carefully to see if its particular structure suggests a new or modified approach. Finally, we should simply pick the best of all the schedules that our analyses have found.

In summary, NP-completeness tells us that a problem may be extremely difficult, but that should not stop us trying to solve it. We must use every weapon at our disposal and do the best that we can.

11.5 REFERENCES AND FURTHER READING

The subject matter of this chapter was drawn heavily from Garey *et al.* (1978), and Garey and Johnson (1979). Weide (1977) and Fisher (1980) provide general surveys of the worst case analysis of heuristic algorithms. Weide (1977) also discusses average performance as well as worst case. Sahni (1977) discusses general approaches for generating approximation schemes.

There are many papers which concentrate on the worst case analysis of scheduling algorithms. Noteable are Graham (1969), Sahni (1976), Garey *et al.* (1978), and Kise, Uno and Kabata (1979) all of whom discuss the problem of scheduling independent jobs on identical machines; Kise, Ibaraki and Mine (1979), who discuss the $n/1//L_{max}$ problem with non-zero ready times; and Gonzalez and Sahni (1978), who consider the flow-shop and general job-shop. Graham *et al.* (1979) refer to many other such studies.

The problem of scheduling independent jobs on identical machines has many extensions. We shall refer to these in the next chapter.

11.6 PROBLEMS

1. Use the longest processing time algorithm on the following problem. There are m machines and $(2m + 1)$ jobs with

$$p_{2k-1} = p_{2k} = 2m - k \quad \text{for} \quad i = 1, 2, \ldots, m.$$

and $p_{2m+1} = m.$

Show that the longest processing time schedule has $C_{max} = 4m - 1$, whereas the schedule formed by using the list $(J_1, J_2, \ldots, J_{m-1}, J_{2m-1}, J_{2m}, J_m, J_{m+1}, \ldots, J_{2m-2}, J_{2m+1})$ has $C_{max} = 3m$. Hence show that $R_A(I) = 4/3 - 1/3m$ is the best possible result for the longest processing time algorithm.

2. Prove Theorem 11.3.

Hint: Let w_k be the time at which the last of k selected jobs completes processing. Show that, if $A_k(I) = w_k$, $A_k(I) = \text{OPT}(I)$. Hence assume $n > k$ and $A_k(I) > w_k$. Let $p^* = $ maximum processing time in the $(n - k)$ unselected jobs. Deduce that

$$\sum_{i=1}^{n} p_i \geq 2(A_k(I) - p^*) + p^*,$$

and hence $$\text{OPT}(I) \geq \frac{1}{2} \sum_{i=1}^{n} p_i \geq A_k(I) - \frac{1}{2} p^*.$$

Next note that at least $k + 1$ jobs have processing time $\geq p^*$. Hence one of the two machines must execute at least $[1 + k/2]$ of the tasks. Hence $\text{OPT}(I) \geq [1 + k/2]p^*$ and thus prove the theorem.

3. For the dynamic programming algorithm developed in the Section 11.2, give explicitly the recursion whereby you would generate the set S from the calculated $T(i, j)$. Hence find an optimal schedule for a six job instance with processing times 5, 8, 9, 2, 6, 3 respectively.

4. Consider an $n/m/F/\bar{F}$ problem. Define a *busy schedule* as one in which at all times from 0 to C_{max} at least one machine is processing an operation. Let A be a heuristic that always produces a busy schedule. Show that $R_A(I) \leq n$ for all instances I. Moreover, by considering the example quoted in Problem 5.8.5, show that this bound may be approached arbitrarily closely.

5. Consider an $n/m/F/\bar{F}$ problem. Define the extended SPT heuristic as follows. Let $P_i = \sum_{j=1}^{m} p_{ij}$, the total processing time for job i. Generate a permutation schedule by ordering the jobs according to non-decreasing P_i. Call this schedule an extended SPT schedule. Letting A be this heuristic, show $R_A(I) \leq m$ for all instances I.

6. Prove Theorem 11.8.

Chapter **12**

A Brief Survey of Other Scheduling Problems

12.1 INTRODUCTION

In a very real sense Chapter 11 completed our study: there are no further theoretical developments in the pages to follow. Essentially this chapter is a brief guide to the literature of scheduling, which extends in many directions beyond the job-shop problem that we have been investigating. We define in greater detail the various forms that precedence constraints may take; introduce the notion of resource constraints; extend the job-shop model in a number of ways; consider various approaches to dynamic and stochastic models; and finally introduce three classes of scheduling problems that have a form distinct from that of the job-shop. Throughout we reference the literature in sufficient detail for those interested to be able to investigate further. They will, however, need to follow up the references made in the papers that we refer to, and further pursue their investigations with the guidance of the *Citation Index*.

12.2 PRECEDENCE CONSTRAINTS AND RESOURCE CONSTRAINTS

In Section 4.1 we defined precedence constraints and discussed some examples of the ways in which they might arise. Generally they limit the choice of schedule by demanding that an operation within one job is completed before a particular operation within a different job is begun. Usually, however, they refer to the last operation within the first job and the first operation within the second job, thus demanding that the first job completes before the second is begun. We begin by discussing this more usual case; for the present precedence constraints will be understood to apply between entire jobs. We begin by classifying the structure that may be present within a set of precedence constraints.

The easiest way to display a set of precedence constraints is through a **network** or a **graph**, examples of which are given in Fig. 12.1(a)–(f). In these the ringed numbers are called **nodes** or **vertices**, and are used to

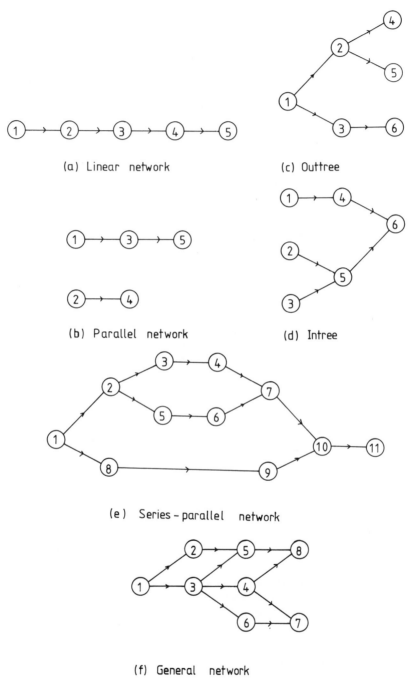

(a) Linear network

(c) Outtree

(b) Parallel network

(d) Intree

(e)   Series – parallel   network

(f) General   network

Fig. 12.1 Some typical examples of precedence networks.

represent jobs. The lines, or **directed edges**, running between some pairs of nodes indicate the precedence constraints. Thus in Fig. 12.1(a) the directed edge between nodes 1 and 2 indicates the constraint that job 1 must complete before job 2 can start.

Figure 12.1(a) illustrates a **linear network** and Fig. 12.1(b) illustrates a **parallel network**. The reason for this nomenclature is obvious. Some authors call a parallel network a **chain network**, but we shall not follow them. Instead we shall use chains in the sense of required strings of jobs; once a chain is initiated all the jobs in it must be completed without interruption by the processing of jobs from another chain.

The **indegree** of a node is the number of directed edges that are incident on it. For example, node 5 in Fig. 12.1(c) has indegree 1 and node 7 in Fig. 12.1(f) has indegree 2. The **outdegree** of the node is the number of directed edges leaving it. For example, node 4 in Fig. 12.1(b) has outdegree 0 and node 3 in Fig. 12.1(f) has outdegree 3. Fig. 12.1(c) and (d) both illustrate a particularly simple form of network called a **tree**. A tree is a connected network in which one, or perhaps both, of the following happens.

Either (i) each node has indegree at most one, in which case the structure
             is called an **outtree** because branching occurs *out* of nodes;
or   (ii) each node has outdegree at most one, in which case the structure
             is called an **intree** because branching occurs *into* nodes.

Thus Fig. 12.1(c) illustrates the outtree and Fig. 12.1(d) illustrates an intree. The linear network in Fig. 12.1(a) is both an intree and an outtree. If a network is composed of more than one tree, it is called a **forest**. Thus the parallel network in Fig. 12.1(b) is both an **outforest** and an **inforest**.

Figure 12.1(e) illustrates a structure known as **series-parallel**. The definition of a series-parallel network is complex. We begin by defining a **parallel subnetwork** to be part of the network consisting of parallel segments all of which share the same set of predecessor nodes and same set of successor nodes. Thus in the figure {3 → 4, 5 → 6} forms a parallel subnetwork. The common set of predecessor nodes is {1, 2} and the common set of successor nodes {7, 10, 11}. A series-parallel network is one that can be reduced to a linear network by iteratively replacing each parallel subnetwork with a linear subnetwork formed by a feasible permutation of the jobs in the parallel subnetwork. Fig. 12.2 shows the steps in reducing Fig. 12.1(e) to a linear network. First we consider the parallel subnetwork {3 → 4, 5 → 6} and replace this by the linear subnetwork {3 → 5 → 4 → 6}. Note that the precedence constraints are maintained. The resulting network is shown in Fig. 12.2(b). {2 → 3 → 5 → 4 → 6 → 7, 8 → 9} now forms a parallel subnetwork and can be reduced leaving the linear network shown in Fig. 12.2(c). This process of iterative reduction to

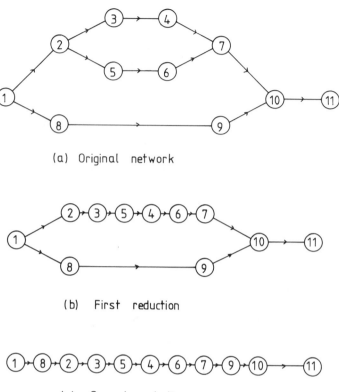

(a) Original network

(b) First reduction

(c) Second reduction

Fig. 12.2 The reduction of the series-parallel network 12.1(e) to the
linear network.

a linear network not only defines series-parallel constraints, but also jus-
tifies their definition. The reduction is often an integral part of polynomial
time algorithms for solving problems with such constraints. (See, for
example, Sidney, 1979.)

Finally there are completely general precedence constraints networks
which fall into none of the above categories. one such is shown in
Fig. 12.1(f).

The importance of this classification is that some problems are NP-hard
with general precedence constraints, but may be solved in polynomial time
when the precedence constraints have a simpler structure. The $n/1//\Sigma \alpha_i C_i$
problem is solvable in polynomial time when tree precedence constraints
exist (Sidney, 1975) or when series-parallel precedence constraints exist
(Lawler, 1978), but not when general precedence constraints exist
(Lenstra and Rinnooy Kan, 1978). The $n/2/F/F_{max}$ problem is solvable in

polynomial time with series-parallel precedence constraints (Sidney, 1979; Monma, 1979), but NP-hard in the case of tree, and hence general, precedence constraints (Rinnooy Kan, 1976). Various authors have explored the frontier between polynomially solvable and NP-complete problems with particular reference to increasing complexity of precedence constraints. (Rinnooy Kan, 1976; Lenstra and Rinnooy Kan, 1978; Graham *et al.*, 1979.)

Returning to the most general case when precedence constraints apply between operations rather than entire jobs, there is somewhat less to report. If a problem is NP-hard with general precedence applying between jobs, it is certainly NP-hard when they apply between operations.

It should be noted that any branch and bound or heuristic method based upon schedule generation techniques can be modified to solve problems with precedence constraints of any form whatsoever. The definition of *schedulable* (Section 10.2) in this case is simply understood to take account of not only the technological constraints, but also the precedence constraints.

Shapiro (1980) has recently discussed a scheduling problem with a much stronger form of precedence relation: **coupling**. He was concerned with the tracking of a number of aircraft by a single radar unit. This must send pulses towards each aircraft and at a certain later time be ready to receive the reflected pulse. In the intervening period it may send or receive pulses concerned with other aircraft provided these allow it to be free to receive the reflected pulse. In other words, these constraints demand not only that one operation precede another, but also that it does so by a specified period.

Throughout this book we have been implicitly assuming that, subject to the satisfaction of precedence and technological constraints being satisfied a machine may process an operation at any time, whether or not other machines are processing other operations. In short, we have assumed that at any time all the machines may be working simultaneously. However, in practice this assumption may be unjustified. There may not be enough operators or enough power for all the machines to be working simultaneously. In scheduling programs upon a computer, some may not be able to run together in a time-sharing mode either because they require the same peripherals or simply because their joint memory requirements exceed the computer's core size. Thus there is much incentive to study problems in which the availability of one or more resource is limited, i.e. **resource-constrained scheduling**.

Many authors have studied this problem. Sahney (1972) has considered a simple two machine problem with only one operator and a cost structure which includes the cost incurred each time the operator is switched between machines. Schrage (1970, 1972) and Stinson *et al.* (1978) discuss branch and bound solutions of general resource constrained scheduling. Kafura and Shen (1977) discuss the particular problem of scheduling programs on a number of computers, each computer having a different core size. General surveys of resource-constrained scheduling may be found in Bennington and McGinnis (1973), Davis (1973), Baker (1974), Coffman (1976), and, with particular emphasis on computer scheduling, Gonzalez (1977). Need it be said that in general resource-constrained scheduling problems are very difficult and usually NP-hard (Blazewicz *et al.*, 1980)?

12.3 VARIATIONS OF THE STATIC JOB-SHOP PROBLEM

In Section 1.4 we made a number of assumptions which limited the structure of the job-shop scheduling problem and as a consequence restricted its ability to model real problems arising in industry and elsewhere. However, as we indicated, we made those assumptions mainly because they led to a straightforward development of the theory; it is possible to solve, albeit with considerable difficulty, many job-shop problems in which one or more of these assumptions do not hold. Here we survey briefly the literature concerned with such relaxation of assumptions. We begin each section of our discussion by restating the appropriate assumption.

   1. *Each job is an entity.* If this assumption is dropped the resulting model allows that some operations within the same job may be processed simultaneously. (N.B. this simultaneous processing has implications for the form of the technological constraints; the operations within a job cannot now be ordered according to a linear network.) Pritisker *et al.* (1971) consider the scheduling to n jobs on m machines, where each job requires one operation on each machine and these may be processed in any order with as much simultaneous processing as required. Holloway and Nelson (1975), and Yoshida and Hitomi (1979) consider the case where overlap between operations is allowed, but completely simultaneous processing is not. In essence, they take the job-shop model very much as we have defined it, but allow that the setting up of a machine for an operation may occur while the previous operation of that job is still processing.

   2. *No pre-emption.* The theory of pre-emptive scheduling has been much studied, particularly in the area of computer scheduling for there it is common to interrupt the processing of one program to expedite that of another. Coffman (1976), Gonzalez (1977), and Graham *et al.* (1979) all survey results in this area. Generally allowing pre-emption makes a problem easier to solve (see, e.g. Baker *et al.*, 1980), although this does not necessarily mean that NP-hard problems suddenly become solvable (Gonzalez and Sahni, 1978). Pre-emption brings two major advantages. Firstly, when choosing an operation to allocate to an idle machine, we need not worry what operations may become available in the near future; we simply

choose from those available now and pre-empt the processing if a better choice becomes available later. Secondly, in using mathematical programming methods we can formulate the problem without the need for integer variables; fractions of an operation can be processed (Labetoulle and Lawler, 1978).

3. *Each job has m distinct operations, one on each machine.* In fact, if you check back through previous chapters you will find that we only needed this assumption when discussing flow-shops. Indeed, in Section 5.4 we explicitly dropped it when discussing Johnson's solution of the $n/2/G/F_{max}$ problem. Moreover, a glance at the schedule generating algorithms of Section 10.2 will show that they do not require this assumption. Thus any heuristic or implicit enumeration method based upon these will solve a job-shop problem regardless of whether each job must be processed once and once only on each machine. In short, we made this assumption purely for tidiness; it made our definitions of the flow-shop and job-shop strictly compatible.

4. *No cancellation.* There are two ways in which a job may be cancelled. Firstly, the scheduler may decide to cancel it and, secondly, a client may cancel the order. The first case is best dealt with by the choice of performance measure. For example, we saw that Moore's algorithm for the $n/1//n_T$ problem (Section 3.5) put jobs that could not meet their due-dates at the end of the processing sequence. If a tardy job is truely worthless, an assumption which underlies the use of n_T, then the scheduler will simply cancel the rejected jobs at the end of the sequence. The second case, in which the client may cancel a job, is completely different; it changes the entire nature of the scheduling problem. Obviously the client is not going to decide to cancel a job at a predetermined time; hence allowing him the option to cancel changes the deterministic scheduling problem into a stochastic one.

5. *The processing times are independent of the schedule.* We have already encountered two ways of solving particular problems in which this assumption does not hold (Problems 6.6.7 and 7.6.7). In general, scheduling problems with sequence dependent processing times are equivalent (i.e. polynomially reducible) to a notorious NP-hard combinatorial optimisation problem known as the **travelling salesman problem** (See Conway, Maxwell, and Miller, 1967). General approaches to this problem are surveyed by Garey and Johnson (1979) and Klee (1980). The following authors all consider this problem specifically in the context of scheduling with sequence dependent set-up times: Baker (1974), Corwin and Esogbue (1974), Lockett and Muhlemann (1972), Rinnooy Kan (1976), and White and Wilson (1977).

6. *In-process inventory is allowed.* The assumption that jobs may wait between machines is unrealistic in many cases. We have already indicated

that in steel mills, for instance, it is important for the processing to be continuous; otherwise the steel will cool and become unworkable. However, a restriction on the number of jobs waiting between machines can arise in other ways. For example, in scheduling programs on a computer there may be insufficient storage to save the output of a program in memory until the physical output device is free. Such problems, in which in-process inventory is not allowed, are called **no-wait** problems and have been much studied: see e.g. Panwalker and Woollam (1979, 1980), Wismer (1972), and surveys of Gonzalez (1977) and Graham *et al.* (1979). The NP-hardness of no-wait problems has been investigated by Sahni and Cho (1979). An intermediate case in which limited storage for in-process inventory is available has been investigated by Papadimitriou and Kanellakis (1980).

7. *There is only one of each type of machine.* Often in practice machines are duplicated and in such cases this assumption is inappropriate. A common problem, particularly in computer scheduling, is to process *n* jobs on *m* machines, when each job requires just one operation and that operation may be performed on any of the machines. The problem that we discussed in Section 11.2 took this form, and this general area of theory is known as **parallel machine or multiprocessor scheduling.** There are three classes of problem depending on whether the machines are **identical, uniform,** or **unrelated.** When the machines are identical, the processing time of J_i is the same on all machines. When the machines are uniform, the processing time varies in a simple fashion: the processing of J_i on M_j takes p_i/s_j where p_i is a constant for J_i and s_j is a **speed factor** associated with M_j. When the machines are unrelated, the processing time of J_i again varies between machines, but in a completely arbitrary fashion. The literature on such problems is enormous; important surveys are given by Coffman (1976), Gonzalez (1977), Graham *et al.* (1979), and Lenstra and Rinnooy Kan (1980).

A flow-shop problem in which each machine may be duplicated has been studied by Salvador (1973). His model also assumes that no in-process inventory is allowed. More general job-shop problems involving duplicated machines are predictably very difficult to solve. Such problems lead, in fact, to a consideration of resource constrained scheduling, which we have already discussed; see Rinnooy Kan (1976).

8. *Machines may be idle.* It is hard to imagine how this assumption can be dropped. However, we may, of course, ensure by means of an appropriate choice of performance measure that the idle time of some or all machines is minimised.

9. *No machine may process more than one operation at a time.* As we have noted this assumption is essentially the same as assumption 7. Consequently, if it is relaxed, we are again led to a study of multiprocessors.

10. *Machines never breakdown and are available throughout the scheduling period.* If we are prepared to admit the possibility of a machine breaking down then our problem becomes stochastic. However it is possible to consider the planned unavailability of a machine during the scheduling period either because it is required to process some job distinct from those to be scheduled or because it is required for routine maintenance. Baker and Nuttle (1980) have considered such problems in the case of single machine processing.

11. *The technological constraints are known in advance and immutable.* This assumption is always made except in one sense. We have considered flow-shops and job-shops. There are also **open shops**. In these each job has to be processed on each machine, but there is no particular order to follow. Thus in open shops the schedule determines not only the order in which machines process the jobs, but also that in which the jobs pass between machines. The open-shop is not as well researched as the flow-shop and the job-shop. For the results that are known see Gonzalez and Sahni (1976), Graham *et al.* (1979), and Sahni and Cho (1979).

12. *There is no randomness; all the data are known and fixed.* When this assumption is relaxed we move into the realm of stochastic scheduling, which we discuss in the next section. We might, however, mention here problems in which the processing times are neither fixed nor random, but instead under the control of the scheduler, who may at a cost make certain machines work faster. See Vickson (1980) and Van Wassenhove and Baker (1980).

As we noted in Section 1.7, the realism of our job-shop model is limited not only by our assumptions about its structure, but also by our choice of performance measures. We have admitted that each of those that we have used represents only a component of a schedule's cost. Yet in practice it is the total cost that we wish to minimise. In restricting our discussion to simple, component cost measures such as C_{max}, $\Sigma w_i T_i$, and n_T we have apparently limited the practical relevance of our studies. However, this really is not so; for we have learned much about the inherent difficulty of scheduling problems, about the structure of implicit enumeration methods, and about heuristic methods. These, and indeed all the concepts and techniques that we have studied, may be applied to develop algorithms that minimise, perhaps approximately, the total scheduling cost. There is a recent and growing literature which seeks to do just this.

We have already referred to the development of algorithms to find schedules that are efficient with respect to a number of criteria and have shown how these may be used to minimise total cost functions that depend on these criteria alone. (N.B. Although we assumed a linear total cost function in Section 4.5, nonlinear functions may be optimised nearly as easily.) However, it is not necessary to develop the set of all efficient schedules in order to minimise a total cost function. In fact, computationally it is usually a very poor approach. The advantage brought by the generation of the efficient set is not simply that the total cost may be minimised, but that management may explore a sensible set of alternative schedules without needing to specify the total cost function precisely. (See Hwang and Masud, 1979.)

A number of approaches which directly minimise a total cost function have been reported. In a sequence of papers Gelders and his co-workers have developed algorithms for a variety of problems in which the sum of weighted flow times and weighted tardinesses, viz. $\sum_{i=1}^{n}(\alpha_i F_i + \beta_i T_i)$, is to be minimised. (Gelders and Kleindorfer, 1974, 1975; Gelders and Sambandam, 1978, and Van Wassenhove and Gelders, 1978). Although this performance measure is clearly not the most general total cost function possible, there is evidence to suggest that it is applicable in many practical cases (Panwalker *et al.*, 1973). Gupta (1975) discusses a search algorithm for minimising a general cost function, but one suspects that it has very poor exponential time complexity. Kao (1980) ingeniously combines branch and bound and dynamic programming in a general algorithm, which minimises a multiattribute cost function.

12.4 DYNAMIC AND STOCHASTIC SCHEDULING

As we remarked in Chapter 1, our model of the job-shop is static and deterministic. We have not considered cases where new jobs are continually arriving; where processing times are uncertain; and where machine, may breakdown. In short, we have not considered dynamic and stochastic job-shops. Nor is it our intention to do so here; we shall simply indicate those areas of literature where the interested reader may search.

It may be argued that uncertainty can enter a problem in two ways. Firstly, it can arise because of our inability to measure anything perfectly. The underlying problem may be deterministic in the sense that all quantities are fixed; however, the values of these quantities may be uncertain, in the sense that they can only be measured with error. The question then arises of how good our scheduling procedures are when they are based upon estimates of processing times, etc., rather than their actual values. Such questions of **robustness** have been investigated by Muth (1963), Conway, Maxwell and Miller (1967, Section 3.6), and Holloway and Nelson (1974b). Secondly uncertainty can arise because the quantities in a problem have an inherent variability. For instance, the processing times of apparently identical operations may vary considerably; we have all met the trivial example of an inspection plate, which may be child's play to remove on one occasion and infuriatingly difficult to remove on another. Most of the papers cited in the following paragraphs are explicitly addressed to the solution of problems with such inherent uncertainty. However, the distinction that we have made between measurement uncertainty and inherent variability is a fine one and methods for treating one form of uncertainty are invariably directly applicable to the other, and to the case where both forms co-exist within the same problem.

The predominant theoretical approach to stochastic scheduling is that of **queuing theory**. In this the jobs are assumed to arrive in a random process, which has a known statistical form. They then queue until their first machine is free, where upon a job is selected from the queue and assigned to that machine according to some predetermined **priority rule or service discipline**. The processing time of the assigned job is assumed to be a random variable with a known distribution. Upon completion either the job queues for processing on its next machine or, if all its processing has been completed, it is discharged from the system. The randomness in the arrival stream, the priority rule, and the randomness in the processing times together imply distributions of flow times for the jobs and idle times for the machines. From these distributions we can deduce the expected cost of a job. Our intention is to find a priority rule that leads to the least expected cost. Unfortunately this is a very difficult task.

First, it is only possible to derive explicit expressions for the expected cost in a limited number of cases when the probability distributions involved have simple and compatible forms. When they do not, queuing theory cannot lead us to the optimal solution of the scheduling problem. Even when they do have suitable forms, deducing the optimal priority rule is very difficult. However it is fairly easy to compare the performance of a number of priority rules, which may have been suggested by experience, and then select the best of these. It is usually in this sense that queuing theory has provided the solution to scheduling problems.

General reviews of queuing theory may be found in Conolly (1975, 1981). Feller (1957), Jaiswal (1968), and Moran (1968). An excellent introduction to the interrelation between queuing theory and stochastic scheduling is given by Conway, Maxwell and Miller (1967). The priority rules referred to here are the direct counterparts of the priority rules used in heuristic schedule generation for deterministic problems. (See Section 10.2). The set of schedulable problems at time t may be viewed as the queue for a machine at time t. Panwalker and Iskander (1977) provide an extensive survey of priority rules in both contexts.

A theoretical approach to stochastic scheduling entirely distinct from queuing theory has recently been developed by Gittins and his co-workers. Their method assigns a **dynamic allocation index** to each job and then schedules the jobs in decreasing order of this index. These indices are updated as the jobs are processed, thus allowing the schedule to adapt to the actual pattern of arrivals and processing times. Gittins and Nash (1977) provide an introductory survey of the major results. Other papers of importance are: Gittins (1979, 1981), Glazebrook (1980), Glazebrook and Nash (1976), and Whittle (1980).

Simulation is probably the technique most used to investigate dynamic and stochastic job-shops. Essentially one models the job-shop on a computer and observes its behaviour under a simulated, that is randomly generated, stream of job arrivals. Because the modern computer is so fast the performance of different scheduling systems may be compared over many 'years'. The literature on simulation in operational research is enormous; most introductory operational research textbooks contain one or two chapters on the subject. Conway, Maxwell and Miller (1967, Chapter 11) provide an excellent introduction to simulation in the context of the job-shop. Recent papers of note are: Arumugam and Ramani (1980); Berry and Finlay (1976); Holloway and Nelson (1974b); Hurrion (1978) and Nelson, Holloway and Wong (1977).

12.5 ASSEMBLY LINE BALANCING; PROJECT MANAGEMENT; AND TIMETABLING OF LECTURES

Not all scheduling problems can be framed within the context of a job-shop. Many have a structure which cannot be described naturally by the processing of jobs on machines. Nonetheless, our study of the job-shop is highly relevant to the discussion and solution of such problems. The same terms and concepts are needed; the same solution methods are required; and, alas, the theory of NP-completeness shows that these problems have the same hardness. Here we describe three important and commonly encountered examples of such scheduling problems.

Assembly Line Balancing. Much of industrial production is based upon an assembly line, along which components are assembled into the finished product. The simplest way to picture this is to imagine a conveyor belt along which are situated a number of **work stations**. At each work station a number of assembly operations are performed. The conveyor belt moves the partially assembled product between work stations; pauses for a length of time to allow the work stations to complete their appointed tasks; and then moves the product along to its next work station. Two observations are apparent. First, each work station is allowed the same time to complete its assembly operations (unless the conveyor belt is peculiarly elastic). Secondly, at equal intervals one unit of the finished product rolls off the assembly line. The length of these intervals is called the **cycle time** and is equal to the time spent at a work station plus the time needed to move between stations.

Assembly line balancing problems may take a number of forms. In each case we are given a list of assembly operations, the time required to perform each, and the precedence constraints that govern the order in which assembly may be completed. Our task is to assign the operations to the work stations. The objective might be to maximise the production rate for a given number of work stations or to minimise the number of work stations required to achieve a given production rate. More generally, we might wish to minimise the unit cost of production, which relates both to the capital investment and the labour costs, the former being partially determined by the number of work stations and the latter both by the number of work stations and by the production rate. Whatever the objective, it must be modified by an important practical consideration. Each work station must be assigned roughly the same amount of work. At least, each manned work station must be; machines do not complain if one has to work harder than another. It is because of this intention to 'balance' the work load evenly along the assembly line that the problem earns its name.

The following authors consider various aspects of assembly line balancing: Bennett and Byrd (1976); Dar-El (1975, 1978); Gehrlein and Patterson (1975, 1978); Kao (1979, 1981); King (1975); Sphicas and Silverman (1976); and Van Assche and Herroelen (1978).

Project Management. Amongst all scheduling techniques probably the best known is that of **critical path analysis** or **CPA**. This is concerned with the scheduling of the many interrelated tasks that together form a complete project. Certain precedence constraints exist between the tasks and the problem is to complete the project as quickly as possible. For instance, in building a house the various tasks include: buying the land, designing the house, ordering the building materials, digging and laying the foundations, building the window frames and doors, building the walls, building the roof, plumbing, wiring, fitting cupboards, decorating, and landscaping the garden. Certain tasks may be executed simultaneously, for example a carpenter may build the window frames and doors while a labourer is laying the foundations. Most tasks can only be begun after others have been completed, for example it is impossible to build the roof until the walls have been finished. The problem is to schedule the various tasks so that the house is built as quickly as possible.

In essence, these critical path problems take the form of a job-shop with an unlimited number of parallel machines, which is required to process a number of single-operation jobs, between which precedence constraints exist. The aim is to minimise make-span. Because machine availability does not limit the number of jobs that may be processed simultaneously, these project management problems have a very simple solution algorithm. Of course, this ability to process an unlimited number of taks simultaneously is very unrealistic in practice and a variety of problems have been studied in which limits are imposed. All are NP-hard.

Project management problems arise in many areas: the construction industry, research and development, logistical planning, etc. The literature on the subject is truly enormous and the following references are by no means fully representative. Baker (1974), Daellenbach and George (1979), Kaufmann and Desbazeille (1969), and Lockyer (1969) all introduce the basic problem and discuss some of its variations. Problems in which the simultaneous processing of tasks is limited are surveyed by Bennington and McGinnis (1973). Other variations of the problem consider the possibility of processing certain tasks faster, but at a cost; see Hindelang and Muth (1979), Kelley (1961), and Moore *et al.* (1978).

Timetabling of Lectures. Although many students will not believe so, it is extremely difficult to design a lecture timetable that allows them the freedom to study all the combination of options that they might wish. The problem is to assign lectures to times on the timetable so that various constraints are satisfied. Obvious constraints are: no lecturer can give two or more lectures at the same time; similarly, compulsory courses cannot be lectured simultaneously or the students would have to be in two places at the same time; each course must have a required number of lectures per week; and so on. Moreover, there are often many less obvious constraints. Lecturers have other duties, e.g. research and administration, and a single lecture period may be too short to undertake any one of these. Thus the timetable should ensure that lecturers' free periods include, perhaps, a complete free afternoon. Equally it is unsatisfactory both for lecturers and students if they are asked to give or attend too many consecutive lectures. Need I go on? It is clear that a very difficult combinatorial problem underlies the construction of a lecture timetable. I have framed this discussion in terms familiar to me, but the same difficulties arise in schools. Indeed, given the existence of split-site schools where a considerable distance may separate the separate buildings, the school timetable problem is compounded with a logistical one. Among the authors who have considered these problems are Almond (1965), Lawrie (1969), Papoulias (1980), Papoulias and Panayiotopoulos (1979) and White (1975).

Hints and Solutions to Problems

1.2 PROBLEMS

See Section 1.8 for full solutions.

1.10 PROBLEMS

2. The first set of relations may be seen in Fig. 1.2. The second set of relations follow immediately on summing over the jobs and dividing by n. (Cf. the proof of Theorem 2.2.)

3. There are three possibilities for L_i: $L_i > 0$, $L_i = 0$, $L_i < 0$. If $L_i > 0$, then $T_i = \max\{L_i, 0\}) = L_i$ and $E_i = \max\{-L_i, 0\} = 0 \Rightarrow T_i - E_i = L_i$. If $L_i = 0$, then $T_i = \max\{0, 0\} = 0$ and $E_i = \max\{0, 0\} = 0 \Rightarrow T_i - E_i = 0 = L_i$. If $L_i < 0$, then $T_i = \max\{L_i, 0\} = 0$ and $E_i = \max\{-L_i, 0\} = -L_i \Rightarrow T_i - E_i = L_i$. So in all cases: $T_i - E_i = L_i$.

4. In the following let $C_1 \leqslant C_1', C_2 \leqslant C_2' \ldots, C_n \leqslant C_n'$.

   (i) If $\quad R(C_1, C_2, \ldots, C_n) = \bar{F}$, we have

$$R(C_1, C_2, \ldots, C_n) = \frac{1}{n} \sum_{i=1}^{n} F_i$$

$$= \frac{1}{n} \sum_{i=1}^{n} (C_i - r_i)$$

$$\leqslant \frac{1}{n} \sum_{i=1}^{n} (C_i' - r_i)$$

$$= R(C_1', C_2', \ldots, C_n').$$

So $\bar{F}$ is a regular measure.

If $R(C_1, C_2, \ldots, C_n) = T_{\max}$, we have

$$R(C_1, C_2, \ldots, C_n) = \max\{T_1, T_2, \ldots, T_n\}.$$

Now $T_i = \max\{L_i, 0\} = \max\{C_i - d_i, 0\} \leqslant \max\{C_i' - d_i, 0\} = T_i'$.

So
$$R(C_1, C_2, \ldots, C_n) = \max\{T_1, T_2, \ldots, T_n\}$$
$$\leq \max\{T'_1, T'_2, \ldots, T'_n\}$$
$$= R(C'_1, C'_2, \ldots, C'_n)$$

Hence T_{max} is regular.

It is an easy matter to show that n_T is regular. By the above $T_i \leq T'_i$ for all jobs. Hence the number of jobs with positive tardiness cannot increase with decreasing completion times.

(ii) $E_i = \max\{-L_i, 0\} = \max\{(d_i - C_i), 0\}$
$$\geq \max\{(d_i - C'_i), 0\}$$
$$= E'_1.$$

So $C_i \leq C'_i \Rightarrow E_i \geq E'_i$. Hence neither $\bar{E}$ nor E_{max} can be non-decreasing in the completion times. Thus neither are regular. It would make sense to minimise $\bar{E}$ or E_{max} whenever the dominant cost is that of storing the finished jobs between their completion and delivery dates.

5. The schedule is feasible. Its Gantt diagram is shown in Figure 10.1.

2.5 PROBLEMS

1. The method of proof is identical to that of Theorem 2.2. Multiply eqn. (2.1) by α_i, sum over i, and the result follows.

2. The total processing time for machine $M_j = \Sigma_{i=1}^n p_{ij}$. Thus the proportion of the make-span for which M_j is actually processing $= \Sigma_{i=1}^n p_{ij}/C_{max}$.

So the mean utilisation $= \dfrac{1}{m} \displaystyle\sum_{j=1}^m \left[\sum_{i=1}^n p_{ij}/C_{max} \right] = \left[\dfrac{1}{m} \sum_{j=1}^m \sum_{i=1}^n p_{ij} \right]/C_{max}.$

Since the term in square brackets is a constant independent of the schedule, it follows that maximising the mean utilisation is equivalent to minimising C_{max}.

3. We have $I_j = C_{max} - \Sigma_{i=1}^n p_{ij}.$

Hence
$$\sum_{j=1}^m \alpha_j I_j = \sum_{j=1}^m \alpha_j \left[C_{max} - \sum_{i=1}^n p_{ij} \right]$$
$$= C_{max} \left(\sum_{j=1}^m \alpha_j \right) - \sum_{j=1}^m \alpha_j \sum_{i=1}^n p_{ij}$$
$$= C_{max} - \sum_{j=1}^m \alpha_j \sum_{i=1}^n p_{ij}, \quad \text{since} \quad \sum_{j=1}^m \alpha_j = 1.$$

Thus $\Sigma_{j=1}^{m}\alpha_j I_j$ is equivalent to C_{max}, which is in turn equivalent to $\bar{I}$ by Theorem 2.4, and the equivalence of $\Sigma_{j=1}^{m}\alpha_j I_j$ and $\bar{I}$ follows immediately.

4. For the schedule which processes J_1 before J_2, $\bar{L} = \bar{T} = 0$. For the schedule which processes J_2 before J_1, $\bar{L} = -\frac{1}{2}$ and $\bar{T} = \frac{1}{2}$. Hence the former schedule is optimal for $\bar{T}$, whilst the latter is optimal for $\bar{L}$.

5. Let $\delta_i(t) = \begin{cases} 1 & \text{if } J_i \text{ unfinished at time } t; \\ 0 & \text{otherwise} \end{cases}$

Thus $\qquad \delta_i(t) = \begin{cases} 1 & \text{if } t \le C_i; \\ 0 & \text{if } t > C_i. \end{cases}$

$$\Rightarrow \int_0^{C_{max}} \delta_i(t)dt = C_i.$$

Now $\qquad\qquad N_u(t) = \sum_{i=1}^{n} \delta_i(t).$

So $\qquad\qquad \int_0^{C_{max}} N_u(t)dt = \sum_{i=1}^{n} C_i = n\bar{C}.$

Hence $\bar{N}_u = n\bar{C}/C_{max}$ and we deduce $\bar{N}_u$ and $(\bar{C}/C_{max})$ are equivalent.

6. $\bar{N}_u = \bar{N}_p + \bar{N}_w.$

So $\quad \bar{N}_w = \bar{N}_u - \bar{N}_p$

$\qquad = \dfrac{n\bar{C}}{C_{max}} - \dfrac{\displaystyle\sum_{i=1}^{n}\sum_{j=1}^{m} P_{ij}}{C_{max}}$ (by solution to problem 5 above and equation (2.3)).

$\qquad = n\bar{W}/C_{max} \qquad$ (by equation (2.2)).

Thus $\bar{N}_w$ and $(\bar{W}/C_{max})$ are equivalent.

7.(a) The first schedule has $C_1 = C_2 = 2$, $(\bar{C}/C_{max}) = 1$ and the second has $C'_1 = 2, C'_2 = 4$, $(\bar{C}'/C'_{max}) = 3/4$. It follows that $(\bar{C}/C_{max})$ is not regular.

(b) The first schedule has $C_1 = 21$, $C_2 = 20$, $C_3 = 21$ and $(\bar{W}/C_{max}) = 3/7$. The second has $C'_1 = 21$. $C'_2 = 30$, $C'_3 = 21$ and $(\bar{W}'/C_{max}) = 37/90 < 3/7$. It follows that $(\bar{W}/C_{max})$ is not regular.

3.7 PROBLEMS

1. Consider the situation shown in Fig. 3.1. Suppose that (a) I is broken into more than two segments and (b) K is itself pre-empted. Confine attention to the initial (leftmost) segment of I. If this segment is interchanged with the first segment of K, the completion times of all jobs are unchanged, and hence so is the value of a regular performance measure. Combining this observation with the sketch proof of Theorem 3.2 given in Section 3.2,

we see that, whether or not the job that pre-empts I is itself pre-empted, we can interchange the initial segment of I and the segment that pre-empts it. Thus we can repeatedly move the initial segment of I to the right until it is contiguous with the second segment of I. Clearly we may now take these two segments to be a single initial segment of I. Again we can move it to the left. Eventually we must succeed in joining the segments of I into a single, continuous whole. At no time have we increased the value of the performance measure. We can repeat this procedure for each pre-empted job. Thus Theorem 3.2 follows in a straight-forward fashion.

2. Consider the alternative proof of Theorem 3.3. Equations (3.2) and (3.3) hold whatever the performance measure. Hence the contribution of I and K to $1/n(\sum_{i=1}^{n} F_i^2)$ in S is

$$\frac{1}{n}(F_I^2 + F_K^2) = \frac{1}{n}((a + p_I)^2 + (a + p_I + p_K)^2),$$

$$> \frac{1}{n}((a + p_K)^2 + (a + p_I + p_K)^2). \quad \text{since} \quad p_I > p_K.$$

$$= \frac{1}{n}((F_K')^2 + (F_I')^2),$$

which is the contribution of I and K to $1/n(\sum_{i=1}^{n} F_i^2)$ in S'. The required result follows.

3. Note first that the flow time of the Kth job in the processing sequence is given by:

$$F_{i(K)} = \sum_{k=1}^{K} p_{i(k)}.$$

Hence
$$\bar{F} = \frac{1}{n}\sum_{i=1}^{n} F_i$$

$$= \frac{1}{n}\sum_{K=1}^{n} F_{i(K)}, \text{ on reordering the sum into processing order.}$$

$$= \frac{1}{n}\sum_{K=1}^{n}\sum_{k=1}^{K} p_{i(k)}$$

$$= \frac{1}{n}\sum_{K=1}^{n}(n + 1 - K)p_{i(K)}.$$

Similarly $W_{i(k)} = \sum_{k=1}^{K-1} p_{i(k)}$; and it follows that $\bar{W} = 1/n \sum_{K=1}^{n}(n - K)p_{i(K)}$.

4. An SPT schedule is $(6, 8, 2, 5, 7, 3, 4, 1)$, for which $\bar{C} = 97/8$ and $C_{max} = 32$. From the solution of Problem 2.5.5 we have that $\bar{N}_u = n\bar{C}/C_{max}$. So here $\bar{N}_u = 97/32$.

5. Use the method of the alternative proof of Theorem 3.3. Thus, briefly, consider a sequence S not in the suggested order with adjacent jobs I and K (I preceding K) such that:

$$\frac{p_I}{\alpha_I} > \frac{p_K}{\alpha_K} \quad \text{or, equivalently,} \quad \alpha_K p_I > \alpha_I p_K.$$

Compare with the schedule S' in which I and K are interchanged (See Figure 3.2). Then the contribution of I and K to $\Sigma \alpha_i F_i$ in S is

$$\frac{1}{n}(\alpha_I F_I + \alpha_K F_K) = \frac{1}{n}(\alpha_I(a + p_I) + \alpha_K(a + p_I + p_K))$$

$$= \frac{1}{n}(\alpha_I(a + p_I) + \alpha_K(a + p_K) + \alpha_K p_I)$$

$$> \frac{1}{n}(\alpha_I(a + p_I) + \alpha_K(a + p_K) + \alpha_I p_K)$$

$$= \frac{1}{n}(\alpha_I(a + p_K + p_I) + \alpha_K(a + p_K))$$

$$= \frac{1}{n}(\alpha_I F_I' + \alpha_K F_K'),$$

their contribution in S'. The optimality of the suggested sequence follows immediately.

If some of the $\alpha_i = 0$, it is clear that their respective jobs J_i do not contribute to the performance measure $\Sigma \alpha_i F_i$. Thus these jobs should be processed last. The optimal sequence is therefore to order those jobs with non-zero α_i so that $p_{i(k)}/\alpha_{i(k)}$ is non-decreasing, and then append the remaining jobs (with $\alpha_i = 0$) in any order.

6. EDD sequence is (2, 6, 4, 7, 3, 1, 5) and the optimal $L_{\max}$ is 25.

7. The proof follows that of Theorem 3.4 closely. Consider a sequence S which is not in order of non-decreasing slack time. Then there is a job I starting at time a and immediately preceding job K such that:

$$(d_I - p_I - a) > (d_K - p_K - a).$$

On considering the schedule S' in which I and K are interchanged we see immediately that

$$L_I < L_K'.$$

Moreover, since $p_K > 0$,

$$L_I < L_I'.$$

Hence $L_I < \min(L_K', L_I')$. Letting L be the minimum lateness of the other

$(n - 2)$ jobs, it follows that

$$\min(L, L_I, L_K) < \min(L, L'_K, L'_I)$$

Hence sequencing in non-decreasing order of slack time maximises the minimum lateness. In other words, scheduling in an apparently sensible manner actually achieves a result that most managers would not find at all desirable.

    8. An optimal schedule is (9, 4, 5, 6, 2, 1, 8, 3, 7, 10), which leads to 1 job being late.

    9. An optimal schedule is (6, 7, 8, 4, 1) with jobs 2, 3 and 5 cancelled.

    10. An optimal schedule is generated by the following algorithm. Throughout U is the set of jobs that have yet to be assigned a position in the processing sequence. t is the time that the job being placed in the kth position of the sequence will start.

Step 1 Set $U = \{J_1, J_2, \ldots, J_n\}$; $t = \min_{i=1}^{n}\{r_i\}$; $k = 1$.

Step 2 Select $J_{i(k)}$ from U such that (a) $r_{i(k)} \leq t$ and (b) $d_{i(k)} \leq d_i$ for all J_i in U such that $r_i \leq t$.

Step 3 Delete $J_{i(k)}$ from U; increase k by 1. If $k > n$, *stop*. Otherwise, go to next step.

Step 4 Set $\tau = \min_{J_i \, in \, U}\{r_i\}$ and reset $t = \max\{(t + 1), \tau\}$. Go to *step 2*.

Condition (a) in step 2 ensures that only jobs that can start at time t are considered and condition (b) ensures that of these the job with earliest due date is selected. Step 4 ensures that t is reset from cycle to cycle in a sensible fashion. There is no point in resetting t to $(t + 1)$ if all the unscheduled jobs have ready times greater than this.

    That this algorithm generates an optimal schedule is proved relatively easily. No job can become ready during the processing of another. Thus of those that may be processed select the one with the earliest due date. The optimality of this selection follows from an interchange argument identical to the proof of Theorem 3.4.

    This algorithm is due to Horn (1974).

4.7 PROBLEMS

    1. An optimal sequence is (7, 11, 5, 1, 2, 4, 8, 10, 3, 6, 9, 12).

    2. Lawler's algorithm gives the optimal schedule (2, 1, 4, 6, 5, 3, 7, 8) with $L_{max} = 2$.

    3. Lawler's algorithm gives the optimal schedule (1, 2, 4, 3) with maximum cost = 4.

    4. Using Lawler's algorithm to solve an $n/1//L_{max}$ problem without precedence constraints, we see immediately that the last position in the

processing sequence will be given to a job with the latest possible due date. Cycling through the algorithm we see that an EDD sequence is built up.

5. N.B. note the parallel between Lawler's and Smith's Algorithms.

Algorithm 4.7 (Lawler)

Step 1 Set $k = n$; $\tau = \sum_{i=1}^{n} p_i$; and set V to be the set of jobs that may be processed last.

Step 2 Find $J_{i(k)}$ in V such that $\gamma_{i(k)}(\tau) \leq \gamma_i(\tau)$ for any job J_i in V.

Step 3 Decrease k by 1; decrease τ by $p_{i(k)}$; delete $J_{i(k)}$ from V; and add to V any job that may now be processed 'last'.

Step 4 If $k \geq 1$, go to *Step 2*. Otherwise, stop with the optimal processing sequence $(J_{i(1)}, J_{i(2)}, \ldots, J_{i(n)})$.

6. Smith's algorithm gives the schedule $(6, 2, 3, 7, 4, 5, 1)$ with $\bar{F} = 110/7$ and $T_{max} = 2$.

7. Using Algorithm 4.6 we find the following sequence of efficient schedules:

$$(6, 2, 5, 3, 1, 7, 4) \text{ with } \bar{F} = 93/7 \text{ and } T_{max} = 11;$$
$$(6, 2, 5, 3, 7, 4, 1) \text{ with } \bar{F} = 98/7 \text{ and } T_{max} = 5;$$
$$(6, 2, 3, 7, 4, 5, 1) \text{ with } \bar{F} = 110/7 \text{ and } T_{max} = 2;$$

and $\quad (2, 3, 7, 4, 6, 5, 1) \text{ with } \bar{F} = 129/7 \text{ and } T_{max} = 1.$

The minimum cost schedule is $(6, 2, 5, 3, 7, 4, 1)$ with total cost $= 108$.

5.8 PROBLEMS

1. Simply draw the Gantt diagrams for the suggested schedules and the results are immediate.

2. The second part of Theorem 5.4 can be proved in a similar fashion to the first part. However, there is an alternative and very informative proof, which we indicate here.

Let $(J_{i(1)}, J_{i(2)}, \ldots, J_{i(n)})$ be any permutation schedule. Then I claim that the make-span is given by

$$\max_{k=1}^{n-1} \left\{ \sum_{k=1}^{K} a_{i(k)} + \sum_{k=K}^{n} b_{i(k)} \right\} \tag{S.1}$$

To see this remember the form of the Gantt diagram for an $n/2/F/F_{max}$ problem. Processing must be contiguous on the first machine; there can be no idle time. However, there may be idle time on the second machine. Let K^* be the smallest number such that $(J_{i(K^*)}, J_{i(K^*+1)}, \ldots, J_{i(n)})$ are processed contiguously on M_2, i.e. without any idle time between jobs. Since K^* is minimal, there must be idle time preceding its processing on M_2. Hence

$J_{i(K^*)}$ starts on M_2 as it completes on M_1 and it follows that

$$F_{max} = \sum_{k=1}^{K^*} a_{i(k)} + \sum_{k=K^*}^{n} b_{i(k)}.$$

We now show that K^* is the maximising k in expression (S.1) above. Firstly suppose that $K^* < n$ and $K^* < K \le n$. Since there is no idle time on M_2 between any of the jobs $(J_{i(K^*)}, J_{i(K^*+1)} \ldots, J_{i(K)})$ and since $J_{i(K^*)}$ starts on M_2 as $J_{i(K^*+1)}$ starts on M_1 we must have

$$\sum_{k=K^*+1}^{K} a_{i(k)} \le \sum_{k=K^*}^{K-1} b_{i(k)}.$$

It follows that

$$\left\{ \sum_{k=1}^{K^*} a_{i(k)} + \sum_{k=K^*}^{n} b_{i(k)} \right\} - \left\{ \sum_{k=1}^{K} a_{i(k)} + \sum_{k=K}^{n} b_{i(k)} \right\}$$

$$= - \sum_{k=K^*+1}^{K} a_{i(k)} + \sum_{k=K^*}^{K-1} b_{i(k)} \ge 0 \qquad (S.2)$$

Similarly suppose that $1 < K^*$ and that $1 \le K < K^*$. Then since there is idle time immediately before the processing of $J_{i(K^*)}$ on M_2, we must have

$$\sum_{k=K+1}^{K^*} a_{i(k)} \ge \sum_{k=K}^{K^*-1} b_{i(k)}$$

It follows that

$$\left\{ \sum_{k=1}^{K^*} a_{i(k)} + \sum_{k=K^*}^{n} b_{i(k)} \right\} - \left\{ \sum_{k=1}^{K} a_{i(k)} + \sum_{k=K}^{n} b_{i(k)} \right\}$$

$$= \left\{ \sum_{k=K+1}^{K^*} a_{i(k)} - \sum_{k=K}^{K^*-1} b_{i(k)} \right\} \ge 0 \qquad (S.3)$$

From (S.2) and (S.3) it is immediate that K^* is the maximising K in expression (S.1) and hence

$$F_{max} = \max_{K=1}^{n} \left\{ \sum_{k=1}^{K} a_{i(k)} + \sum_{k=K}^{n} b_{i(k)} \right\}. \qquad (S.4)$$

Thus the $n/2/F/F_{max}$ problem is to find the permutation that minimises this expression. But consider the 'reverse' problem in which the machine times are interchanged, i.e. the M_1 times are now M_2 times and *vice versa*. It follows from (S.4) the solution to this reverse problem is simply the reverse of the permutation which optimises the original problem. Now look at

Theorem 5.4. Part (ii) is simply a statement of part (i) for the *reverse problem*. The truth of part (ii) is thus immediate.

3. Johnson's Algorithm gives the schedule $(J_7, J_1, J_2, J_6, J_8, J_4, J_5, J_3)$ with $F_{max} = 49$.

4. Johnson's algorithm must place J_1 first in the schedule. This leads to a value of $\bar{F} = 237/24$. *Any* schedule with J_1 later in the sequence has a strictly smaller value of $\bar{F}$. (See the solution to Problem 5.8.5 below.)

5. Let J_1 be placed Kth in the processing sequence. From the relative magnitudes of β and ε it follows that

$$F_1 \quad = \varepsilon + \tfrac{3}{4}\varepsilon$$
$$F_2 \quad = 2\varepsilon + \tfrac{3}{4}\varepsilon$$

$$\vdots$$

$$F_{K-1} = (K-1)\varepsilon + \tfrac{3}{4}\varepsilon$$
$$F_K \quad = K\varepsilon + \beta$$
$$F_{K+1} = K\varepsilon + \beta + \tfrac{3}{4}\varepsilon$$
$$F_{K+2} = K\varepsilon + \beta + 2.\tfrac{3}{4}\varepsilon$$

N.B. since $\beta > n\varepsilon$ there can be no idle time on M_2 after F_K has been processed.

$$\vdots$$

$$F_n \quad = K\varepsilon + \beta + (n-K)\tfrac{3}{4}\varepsilon$$

$$\Rightarrow \sum_{i=1}^{n} F_n = \frac{K(K-1)}{2}\varepsilon + (K-1)\tfrac{3}{4}\varepsilon + (n-K+1)(K\varepsilon + \beta)$$

$$+ \frac{(n-K)(n-K+1)}{2}\tfrac{3}{4}\varepsilon$$

Now Johnson's Algorithm puts J_1 in the first position $\Rightarrow K = 1$ and

$$\bar{F}_{\text{Johnson}} = \frac{1}{n}\left\{0 + 0 + n(\varepsilon + \beta) + \frac{n(n-1)3\varepsilon}{8}\right\}$$

$$= \varepsilon + \beta + \frac{(n-1)3\varepsilon}{8} \text{ as required.}$$

Simplifying the expression for $\sum_{i=1}^{n} F_n$ gives:

$$\sum_{i=1}^{n} F_i = K^2(-\varepsilon/8) + K(7\varepsilon/8 - \beta + n\varepsilon/4) + \text{terms independent of } K$$

Letting K take non-integral values for the present and differentiating gives

$d/dK(\Sigma_{i=1}^{n} F_i) = -K\varepsilon/4 + (7\varepsilon/8 - \beta + n\varepsilon/4)$. For $K = n$

$$\frac{d}{dK}\left(\sum_{i=1}^{n} F_i\right) = \frac{7\varepsilon}{8} - \beta < 0 \text{ since } n\varepsilon < \beta \Rightarrow \varepsilon < \beta.$$

For $K = 0$

$$\frac{d}{dK}\left(\sum_{i=1}^{n} F_i\right) = \varepsilon(n/4 + 7/8) - \beta < 0 \text{ since } n\varepsilon < \beta$$

$\Rightarrow (\Sigma_{i=1}^{n}F_i)$ is minimised on $0 \leqslant K \leqslant n$ for $K = n$. It follows that any optimal schedule for the $n/2/F/\bar{F}$ problem puts J_1 last in the processing sequence. Hence

$$\bar{F}_{\text{opt}} = \beta/n + (n + 1)\frac{\varepsilon}{2} + \frac{3(n - 1)\varepsilon}{4n}$$

Letting $\varepsilon \to 0 \Rightarrow \bar{F}_{\text{Johnson}}/\bar{F}_{\text{opt}} \to n$.

6. An optimal schedule is
$$M_1 \ (8,13,5,3,1,9,4,14,10,2,6)$$
$$M_2 \ (10,2,6,7,11,12,8,13,5,3,1,9)$$

and this has $F_{\text{max}} = 115$.

7. An optimal schedule is $(5, 6, 2, 3, 4, 1)$ with $F_{\text{max}} = 66$.

8. The hint in the question is substantial. The recursions for s_{i1}, s_{i2} and s_{i3} are straight-forward. (No job can wait for processing on M_2 because $\min\{\alpha_i\} \geqslant \max\{\beta_i\}$.) Now $t_{i1} = \Sigma_{l=1}^{i-1}(\alpha_l + \beta_l) = s_{i1} + \Sigma_{l=1}^{i-1}\beta_l$. It remains to show that $t_{i2} = s_{i3} + \Sigma_{l=1}^{i-1}\beta_l$. Note that for $i = 1$ this is true viz. $t_{12} = \alpha_1 + \beta_1 = s_{11}$. (N.B. $\Sigma_{l=1}^{0}\beta_l = 0$ by convention). Assume inductively that for $i < n$

$$t_{i2} = s_{i3} + \sum_{l=1}^{i-1} \beta_l.$$

Then $\quad t_{(i+1)2} = \max\{t_{i2} + (\beta_i + \gamma_i), t_{(i+1)1} + \alpha_{i+1} + \beta_{i+1}\}$

$$= \max\left\{s_{i3} + \sum_{l-1}^{i} \beta_l + \gamma_i, s_{(i+1)1} + \sum_{l=1}^{i} \beta_l + \alpha_{i+1} + \beta_{i+1}\right\}$$

$$= \max\{s_{i3} + \gamma_i, s_{(i+1)1} + \alpha_{i+1} + \beta_{i+1}\} + \sum_{l=1}^{i} \beta_l$$

$$= s_{(i+1)3} + \sum_{l=1}^{i} \beta_l.$$

Thus $\quad \left(t_{i2} = s_{i3} + \sum_{l=1}^{i-1} \beta_l\right) \text{ implies } \left(t_{(i+1)2} = s_{(i+1)3} + \sum_{l=1}^{i} \beta_l\right)$

and the general result follows inductively from the truth of the case when $i = 1$.

9. Figure 5.7 illustrates the schedule:

$$M_1 \ (J_2, J_1)$$
$$M_2 \ (J_1, J_2)$$
$$M_3 \ (J_2, J_1)$$
$$M_4 \ (J_2, J_1)$$
$$M_5 \ (J_2, J_1)$$
$$M_6 \ (J_2, J_1).$$

The schedule line O A B C D E F G H I J X has an interpretation similar to that given for Fig. 5.6. Here, for example, the segment FG corresponds to M_5 completely processing J_2 while M_3 continues to process J_1. This schedule need not be considered because there are different processing sequences on M_1 and M_2. (See Theorem 5.1).

10. Quite simply, if it is optimal to take the path C D B, it is also optimal to take the path C A B; because by the geometry of a parallelo-gram they have the same length.

11. Drawing the schedule graph indicates clearly that it is only worth considering the two permutation schedules: (a) all machines process J_1 then J_2, or (b) all machines process J_2 then J_1. The former has least total length of vertical segments (and, equivalently, the least total length of horizontal segments), so it is optimal.

6.6 PROBLEMS

1. An optimal schedule is (J_1, J_2, J_3, J_4) with $\bar{T} = 7$.

2. An optimal schedule is (J_1, J_4, J_2, J_3) with $\Sigma \gamma_i = 69$.

3. A little thought shows the dynamic programming recursion (6.3) leads to precisely the first argument by which we deduced Theorem 3.3 in Section 3.3.

4. You should obtain the same answer as by forward dynamic prog-ramming!

5. Applying the suggested algorithm to the problem leads to the schedule (J_2, J_1, J_3) with $\Sigma \gamma_i = 3$. But the schedule (J_1, J_2, J_3) has $\Sigma \gamma_i = 1$.

6. Firstly note that the maximum of $\gamma_{i(k)}(C_{i(k)})$ obeys a similar decom-position to their sum, viz. compare (6.1) with:

$$\max_{k=1}^{n}\{\gamma_{i(k)}(C_{i(k)})\} = \max\{\lambda_K, \gamma_{i(K+1)}(C_{i(K+1)}), \gamma_{i(K+2)}(C_{i(K+2)}), \ldots, \gamma_{i(n)}(C_{i(n)})\}$$

where

$$\lambda_K = \max_{k=1}^{K}\{\gamma_{i(k)}(C_{i(k)})\}.$$

Thus we should ensure that λ_K is as small as possible for each $K = 1, 2,$ $\ldots, n$. Hence as before we may deduce that the first K jobs are in an optimal sequence for the reduced problem. In consequence we deduce the following recursion for $\Gamma(Q)$, the minimum cost of scheduling the jobs in Q:

$$\Gamma(Q) = \begin{cases} \gamma_i(p_i) & \text{if } Q = \{J_i\} \\ \min_{J_i \text{ in } Q}\{\max\{\Gamma(Q - \{J_i\}), \gamma_i(C_Q)\}\} \end{cases}$$

The rest of the problem is purely computational. The optimal sequence is (J_3, J_2, J_1, J_4) with $L_{\max} = 11$.

7. Using the dynamic programming recursions developed in the question, we find (after considerable computation)

$$\Gamma(\{J_1, J_2, J_3, J_4\}, 1) = 15$$
$$\Gamma(\{J_1, J_2, J_3, J_4\}, 2) = 12$$
$$\Gamma(\{J_1, J_2, J_3, J_4\}, 3) = 17$$
$$\Gamma(\{J_1, J_2, J_3, J_4\}, 4) = 13$$

The minimum 12 shows that J_2 should be last in the processing sequence. Looking back through our earlier calculations we find

$$\Gamma(\{J_1, J_2, J_3, J_4\}, 2) = \min\{\Gamma(\{J_1, J_3, J_4\}, 1) + \pi_{12},$$
$$\Gamma(\{J_1, J_3, J_4\}, 3) + \pi_{32}$$
$$\Gamma(\{J_1, J_3, J_4\}, 4) + \pi_{42}\}$$
$$= \min\{5 + 9, 8 + 4, 7 + 6\} = 12.$$

The minimum occurs with J_3 last of J_1, J_3, J_4, so we now know that the optimal sequence has the form $(-, -, J_3, J_2)$. Next we note from our earlier calculations that

$$\Gamma(\{J_1, J_3, J_4\}, 3) = \min\{\Gamma(\{J_1, J_4\}, 1) + \pi_{13},$$
$$\Gamma(\{J_1, J_4\}, 4) + \pi_{43}\}$$
$$= \min\{9 + 9, 6 + 2\} = 8.$$

This minimum occurs with J_4 last of $\{J_1, J_4\}$. Hence we see that the optimal sequence to minimise $C_{\max}$ is (J_1, J_4, J_3, J_2) with $C_{\max} = 12$.

8. To prove (i) put $x = 1$.
   To prove (ii) differentiate with respect to x and put $x = 1$.

9. An optimal schedule is $(J_1, J_4, J_2, J_5, J_6, J_3)$ with $T = 23/6$.

10. Let S be a schedule in which J_k precedes J_i, perhaps with intervening jobs. Simply interchange J_i and J_k. Because $p_i < p_k$ and $d_i < d_k$ no job can become more tardy. The result follows.

11. We may deduce that an optimal schedule exists with J_2 preceding J_1

preceding J_3 and also with J_2 preceding J_4 preceding J_3, but no relation between J_1 and J_4, viz

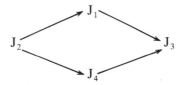

Obviously there are now only two candidates for an optimal schedule, viz (J_2, J_1, J_4, J_3) and (J_2, J_4, J_1, J_3), and it would be simplest to choose between these by direct comparison. However, even using dynamic programming the solution is very fast. For 3 of the 4 columns of Table 6.1 are eliminated; 10 of the 12 in Table 6.2 are eliminated; 10 of the 12 in Table 6.3 are eliminated; and 3 of the 4 in Table 6.4 are eliminated.

7.6 PROBLEMS

1. A straightforward example: you should find the optimal schedule 2134.

2. The optimal schedule is 3412 with minimal $C_{max} = 49$.

3. As we noted in Chapter 5, the optimal schedule for a flow-shop with four or more machines may not be a permutation schedule.

Developing the theory of Section 7.3, we are led to the lower bound

$$lb(A) = \max\{\alpha_{i(K)} + \sum_{J_i \text{ in } U} a_i + \min_{J_i \text{ in } U}\{b_i + c_i + d_i\},$$

$$\beta_{i(K)} + \sum_{J_i \text{ in } U} b_i + \min_{J_i \text{ in } U}\{c_i + d_i\},$$

$$\gamma_{i(K)} + \sum_{J_i \text{ in } U} c_i + \min_{J_i \text{ in } U}\{d_i\},$$

$$\delta_{i(K)} + \sum_{J_i \text{ in } U} d_i\}.$$

Applying these lower bounds to the numerical example shows that 4321 is an optimal schedule with minimal $C_{max} = 36$.

4. The introduction of non-zero ready times means simply that the calculation of $\alpha_{i(k)}$ given in expression (7.1) has to be modified to

$$\alpha_{i(k)} = \begin{cases} r_{i(1)} + a_{i(1)} & \text{for } k = 1 \\ \max\{r_{i(k)}, \alpha_{i(k-1)}\} + a_{i(k)} & \text{for } k > 1. \end{cases}$$

Using this calculation of $\alpha_{i(k)}$ in the branch and bound method of Section 7.3 leads to the optimal schedule 2341 with $C_{max} = 63$. Note that a considerable part of the tree has to be searched because in the early stages the lower bounds (7.5) is poor. Only when the effect of the ready time has been completely absorbed in the $\alpha_{i(k)}$ does elimination occur as fast as we might hope. Hence we are encouraged to modify the lower bound to include some dependence on the ready times of the remaining jobs. For example, we might replace the first component in the lower bound given in (7.2) to

$$C_{max} > \max\{\alpha_{i(K)}, \min_{J_i \text{ in } U}\{r_i\}\} + \sum_{J_i \text{ in } U} a_i + \min_{J_i \text{ in } U}\{b_i + c_i\}.$$

5. The expressions q_2' and q_3', which replace q_2 and q_3 respectively, simply note that there might of necessity be idle time on either of machines M_2 and M_3 after job $J_{i(K)}$. An optimal schedule for the problem is 4321 with $C_{max} = 96$. In your solution, notice how much poorer the components of the lower bounds derived from the second and third machine would be if q_1' and q_2' had not replaced q_1 and q_2.

6. Consider any job J_i in U and ignore any possibility of J_i waiting between the machines. Then the continuous processing of J_i will take $(a_i + b_i + c_i)$. Next consider the other jobs in U and assume that whatever order they are processed in there is no idle time on the third machine. Then

$C_{max} = \alpha_{i(K)}$ + (processing time on M_1 for those jobs in U processed before J_i)

$\qquad$ + (processing of J_i on all three machines)

$\qquad$ + (processing time on M_3 for those jobs in U processed after J_i)

$$= \alpha_{i(K)} + \sum_{J_j \text{ before } J_i} a_j + (a_i + b_i + c_i) + \sum_{J_j \text{ after } J_i} c_j.$$

Assigning the other jobs in U so that the two sums are minimised and remembering that there might be idle time or waiting time, we have

$$C_{max} > \alpha_{i(K)} + (a_i + b_i + c_i) + \sum_{\substack{J_j \text{ in } U \\ J_j \neq J_i}} \min\{a_j, c_j\}.$$

Noting that this is true for every J_i in U, we find the required lower bound.

7. The derivation of the lower bound is trivial: at the node $i_1, i_2, \ldots, i_K$ XXX . . X there are $(n - K)$ jobs still to be assigned and each must take at least π_{min} to process. An optimal schedule for the problem is 1432 with $C_{max} = 12$.

8. Johnson's schedule is 152634 with $C_{max} = 87$. Evaluating lower bounds at nodes according to expression (7.5) gives:

$$lb(1XXXXX) = 86$$
$$lb(2XXXXX) = 88$$
$$lb(3XXXXX) = 90$$
$$lb(4XXXXX) = 90$$
$$lb(5XXXXX) = 87$$
$$lb(6XXXXX) = 88$$

Thus the first trial schedule given by Johnson's algorithm eliminates all branches below nodes 2XXXXX, 3XXXXX, 4XXXXX, 5XXXXX, 6XXXXX. Since $lb(1XXXXX) = 86$, there is a possibility of the branches below this node leading to a better schedule than the trial. However, since $86 \times 1.15 > 87$, the best possible cannot improve on the trial by as much as 15%. Hence we accept 152634 as a satisfactory schedule.

9. The problem of finding good lower bounds for the $n/1//\bar{T}$ problem is a very difficult one. (See Rinnooy Kan, Lageweg and Lenstra, 1975). However, a simple lower bound at $J_{i(1)}J_{i(2)} \ldots J_{i(K)}$ XX .. X is the contribution to $\bar{T}$ from the jobs $J_{i(1)}J_{i(2)} \ldots, J_{i(K)}$ plus the contribution to $\bar{T}$ of all the jobs in U if they are processed next, viz.

$$lb(A) = \frac{1}{n}\left[\sum_{k=1}^{K} T_{i(k)} + \sum_{J_i \text{ in } U} \max\{C_{i(K)} + p_i - d_i, 0\}\right]$$

where $C_{i(K)}$ = completion time of $J_{i(K)}$

$$= \sum_{k=1}^{K} p_{i(k)}$$

The optimal schedule is, of course, as found in Section 6.2: 2143 with $\bar{T} = 5$.

8.4 PROBLEMS

1. Minimise $I_{31} + I_{32} + (X_{11}6 + X_{12}9 + X_{13}4) + (X_{11}8 + X_{12}5 + X_{13}8)$

subject to
$$X_{11} + X_{21} + X_{31} = 1;$$
$$X_{12} + X_{22} + X_{32} = 1;$$
$$X_{13} + X_{23} + X_{33} = 1;$$
$$X_{11} + X_{12} + X_{13} = 1;$$
$$X_{21} + X_{22} + X_{23} = 1;$$
$$X_{31} + X_{32} + X_{33} = 1;$$

$$I_{11} + (X_{12}6 + X_{22}9 + X_{32}4) + W_{12}$$
$$- W_{11} - (X_{11}8 + X_{21}5 + X_{31}8) - I_{21} = 0;$$
$$I_{21} + (X_{12}8 + X_{22}5 + X_{32}8) + W_{22}$$
$$- W_{21} - (X_{11}3 + X_{21}2 + X_{31}17) - I_{31} = 0;$$
$$I_{12} + (X_{13}6 + X_{23}9 + X_{33}4) + W_{13}$$
$$- W_{12} - (X_{12}8 + X_{22}5 + X_{32}8) - I_{22} = 0;$$
$$I_{22} + (X_{13}8 + X_{23}5 + X_{33}8) + W_{23}$$
$$- W_{22} - (X_{12}3 + X_{22}2 + X_{32}17) - I_{32} = 0;$$
$$I_{11} = I_{12} = 0, I_{21} \geq 0, I_{22} \geq 0, I_{31} \geq 0, I_{32} \geq 0;$$
$$W_{11} = W_{21} = 0, W_{12} \geq 0, W_{22} \geq 0, W_{13} \geq 0, W_{23} \geq 0;$$
and $X_{ij} = 0$ or 1 for $i = 1, 2, 3$, and $j = 1, 2, 3$.

2. The flow time of the Kth job is given by

$$F_{i(K)} = \sum_{k=1}^{K} p_{i(k)1} + (W_{1K} + p_{i(K)2}) + (W_{2K} + p_{i(K)3}) + \ldots + (W_{(m-1)K} + p_{i(K)m})$$

$$= \sum_{k=1}^{K} \sum_{i=1}^{n} X_{ik}p_{i1} + \sum_{j=1}^{m-1} W_{jK} + \sum_{j=2}^{m} p_{i(K)j}.$$

Whence

$$\sum_{K=1}^{n} F_{i(K)} = \sum_{K=1}^{n} \left\{ \sum_{k=1}^{K} \sum_{i=1}^{n} X_{ik}p_{i1} + \sum_{j=1}^{m-1} W_{jk} \right\} + \sum_{K=1}^{n} \sum_{j=2}^{m} p_{i(K)j}.$$

Note that the double sum of $p_{i(K)j}$ over K and j is independent of the sequence and it follows that minimising $\bar{F}$ is equivalent to minimising

$$\sum_{K=1}^{n} \left\{ \sum_{k=1}^{K} \sum_{i=1}^{n} X_{ik}p_{i1} + \sum_{j=1}^{m-1} W_{jK} \right\}.$$

9.6 PROBLEMS

1. Remember that all data are integral. Let x be any integer. Then to represent x to base 10 requires $[\log_{10}x + 1]$ digits and to represent it in binary requires $[\log_2 x + 1]$. Now $\log_2 x = \log_2 10 . \log_{10}x$ and $\log_2 10 > 1$. So we have

$$[\log_2 x + 1] = [\log_2 10 . \log_{10}x + 1]$$

$$< \log_2 10 [\log_{10}x + 1]$$

It follows that $v \leqslant (\log_2 10)\eta$.

Similarly, since $\log_{10}x \leqslant \log_2 x$,

$$\eta \leqslant v.$$

2. To encode an $n/1//n_T$ problem we represent numbers to the base 10 first listing the n due dates, then the n processing times; commas are used as the separators. The problem size of the $10/1//n_T$ example is 46.

We may use precisely the same convention to encode an $n/1//\bar{T}$ problem. Thus the $4/1//\bar{T}$ example of Problem 6.6.1 has size 21.

3. In encoding an $n/m/A/B$ problem at least one digit must be used for each of the mn processing times. Thus $v \geqslant mn$ and it follows immediately that

$$v! \geqslant (mn)!$$
$$= [mn.(mn-1)\ldots(mn-n+1)] [(mn-n)\ldots(mn-2n+1)]\ldots$$
$$\ldots[n.(n-1)\ldots 1]$$
$$\geqslant (n!)(n!)\ldots(n!)$$
$$= (n!)^m.$$

4. The proofs of three parts are elementary. For example:
   (i) $f(v)$ is order $O(v^n) \Rightarrow f(v)/v^n \to c_1$ as $v \to \infty$;
   $g(v)$ is order $O(v^m) \Rightarrow g(v)/v^m \to c_2$ as $v \to \infty$.

Hence $f(v)g(v)/v^{m+n} \to c_1 c_2$ as $v \to \infty$ and so $f(v)g(v)$ is $O(v^{m+n})$.

5. $(a_n v^n + a_{n-1} v^{n-1} + \ldots a_0)/v^n = a_n + a_{n-1}/v + \ldots + a_0/v^n$
$$\to a_n \text{ as } v \to \infty.$$

Hence $(a_n v^n + a_{n-1} v^{n-1} + \ldots + a_0)$ is $O(v^n)$.

6. The exact time complexity function that you obtain depends on your exact statement of these algorithms. However in both cases you should find the time complexity if $O(n^2)$ — or less, if you are familiar with techniques of efficient algorithm construction.

N.B. as to be shown in Problem 10.7.11, Johnson's method is equivalent to a simple sort algorithm.

7. Let the algorithm have polynomial time complexity $r(v)$ with respect to the first encoding convention. Consider a problem of size η under the second convention. Its size is at most $q(\eta)$ under the first convention, so the time complexity is at most $r(q(\eta))$, i.e. a polynomial in η.

8. Suppose that $\Pi_1 \propto \Pi_2$ and $\Pi_2 \propto \Pi_3$. Then each instance of Π_1 of size v can be transformed in polynomial time, say $p(v)$, to an instance of Π_2 of size at most $\eta = p(v)$. Since $\Pi_2 \propto \Pi_3$, this constructed instance of Π_2 may be transformed in polynomial time, say $q(\eta) = q(p(v))$ to an instance of size at most $q(\eta) = q(p(v))$. Thus composing these two transformations gives a reduction of the initial instance of Π_1 to an instance of Π_3 in time $[p(v) + q(p(v))]$; that is in polynomial time. Hence $\Pi_1 \propto \Pi_3$.

9. It is trivial to show that $\sim$ is reflexive, symmetric, and transitive. Hence it is an equivalence relation. To show that P is an equivalence class we must show that for all Π_1, Π_2 in P $\Pi_1 \sim \Pi_2$. To show this it is sufficient that for any Π_1 Π_2 in P $\Pi_1 \propto \Pi_2$. Now since Π_1 is in P it may be solved in polynomial time. Suppose we have two instances of Π_2, one giving a *yes* answer and one giving a *no* answer. Reduce any instance of Π_1 to an equivalent instance of Π_2 as follows. First, answer the instance Π_1 in polynomial time. if the answer is *yes*, select as the equivalent instance of Π_2 the one above that gives a *yes* answer. Otherwise select the instance of Π_2 that gives the *no* answer. Since the Π_1 instance can be solved in polynomial time, this reduction takes a little longer and thus $\Pi_1 \propto \Pi_2$. That all NP-complete problems are equivalent is a trivial deduction from the definition of NP-completeness.

10. Suppose that we can answer any instance of size v in polynomial time $p(v)$. First note that $\bar{F} \leq 2^v$ for all schedules since all data $\leq 2^v$ (Remember all numbers are represented in binary.) Now pose the recognition problem which asks whether there is a schedule with $\bar{F} \leq 2^v/2 = 2^{v-1}$. If there is, we know that the optimal $\bar{F} \leq 2^{v-1}$ and in this case we next ask where there is a schedule with $\bar{F} \leq 2^v/4 = 2^{v-2}$. If there was no schedule

with $\bar{F} \le 2^{\nu-1}$, we know that the optimal $\bar{F} < 2^{\nu-1}$ and in this case we next ask whether there is a schedule with $\bar{F} \le \frac{3}{4}.2^{\nu} = 3.2^{\nu-2}$. We carry on this repeated halving of the interval. After k repetitions we will have located the optimal $\bar{F}$ within an interval $2^{\nu-k}$ long. Now since all the data is integral any interval less than $1/n$ long determines the optimal $\bar{F}$ exactly. Noting that $1/\nu \le 1/n$, we are thus guaranteed to find the optimal $\bar{F}$ in $[\nu + \log_2\nu + 1]$ steps. The total number of operations is thus $O([\nu + \log_2\nu + 1]p(\nu)) = O(\nu p(\nu))$ since one recognition problem is solved at each step. Thus the optimisation problem is solved in polynomial time. [Strictly speaking, we might argue that this algorithm does not *solve* the $n/m/G/\bar{F}$ optimisation problem fully since it only finds the minimal $\bar{F}$; it does not find an optimal schedule. However, we shall leave that subtlety to others. For the purposes of Chapter 9 you should note two points. Firstly, if the optimisation problem can be solved in polynomial time, then the recognition problem can be answered in polynomial time also. Secondly and conversely, if the recognition problem cannot be answered in polynomial time, then the optimisation problem cannot be solved in polynomial time either.]

10.7 PROBLEMS

1. The proof is identical to that of Theorem 2.1.

2. Let S be a non-delay schedule. Suppose S is not active. Then there is an operation which could be started without delaying any other operation or violating the technological constraints. Thus under S some machine must be unnecessarily idle. Hence S cannot be a non-delay schedule, which is a contradiction. Thus all non-delay schedules are active.

3. Suppose there are K machines which are candidates for M^*, i.e. machines on which there is a schedulable operation which completes at ϕ^*. Consider the $(K - 1)$ machines not selected for M^*. No operation on these may be considered at step 3 and thus these operations remain schedulable at the next stage. Thus at the next stage ϕ^* is equal to its value at the present stage and the $(K - 1)$ are again candidates for M^*. It follows that over the next $(K - 1)$ stages each of these will be selected for M^* and that the order in which they are selected is immaterial to the final constructed schedule.

4. To show that any active schedule can be generated by Algorithm 10.1 we show the converse: namely, that any schedule that cannot be generated cannot be active. First, note that, if we remove steps 2 and 3, any schedule can be built by a suitable selection of o_j from the entire set of schedulable operations. Consider a schedule R which cannot be generated by the algorithm. Then at some stage the algorithm selects a machine M^* and an operation o_j to process next on M^* according to step 3, whereas in R

the next operation o_k on M^* fails condition (ii) of step 3, i.e. $\sigma_k \geq \phi^*$. Thus in R o_k is scheduled before o_j, yet o_j can quite clearly be leapfrogged over o_k in the processing without causing any delay to o_k. Thus R cannot be active.

5. The argument which shows that Algorithm 3.2 always generates a non-delay schedule is similar to that by which we have shown Algorithm 3.1 always generates an active schedule. The difference between them is that we must remark that, if step 3 of algorithm 3.2 does not select o_j such that $\sigma_j = \sigma^*$, then there must be idle time on M^* and so the generated schedule cannot be non-delay. To show that any non-delay schedule can be generated by Algorithm 3.2 we use an argument similar to that given above in solution to Problem 10.7.4.

6. The schedule that you obtain depends on your choice of priority rules.

7. The schedule 123465 may be obtained with $C_{max} = 86$. In fact the tree shows this to be an optimal schedule. Note that in restricted flooding it may not be necessary to evaluate *all* the lower bounds at the nodes immediately below the one from which all branches are being explored. Here, for example, node 1XXXXX has a lower bound of 86 and 12XXXX, the first node evaluated at the next level down, also has a lower bound of 86. Immediately we know that nodes 13XXXX, 14XXXX, 15XXXX, and 16XXXX cannot have an improved lower bound so no calculations need be performed at these nodes.

8. The six possible schedules are

$$(1, 2, 3) \text{ with } \bar{T} = 4/3$$
$$(1, 3, 2) \text{ with } \bar{T} = 5/3$$
$$(3, 1, 2) \text{ with } \bar{T} = 4/3$$
$$(3, 2, 1) \text{ with } \bar{T} = 5/3$$
$$(2, 3, 1) \text{ with } \bar{T} = 4/3$$
$$(2, 1, 3) \text{ with } \bar{T} = 3/3$$

Starting $(3, 1, 2)$ adjacent pairwise interchange can lead to schedules $(1, 3, 2)$ and $(3, 2, 1)$ both with higher values of $\bar{T}$. Thus the neighbourhood search stops with $(3, 1, 2)$ as the selected best schedule. Yet $(2, 1, 3)$ is clearly optimal.

9. In finding the following solution I have combined steps 2 and 3 of Algorithm 10.3, selecting as the next seed the first improvement found.

Seed $(1,2,3,4)$ $\bar{T} = 7$

Neighbourhood generated:

$(2,1,3,4)$ $\bar{T} = 23/4$ improvement

Seed $(2,1,3,4)$ $\bar{T} = 23/4$
Neighbourhood generated:
         $(1,2,3,4)$ $\bar{T} = 7$
         $(3,1,2,4)$ $\bar{T} = 34/4$
         $(4,1,3,2)$ $\bar{T} = 8$
         $(2,3,1,4)$ $\bar{T} = 25/4$
         $(2,4,3,1)$ $\bar{T} = 6$
         $(2,1,4,3)$ $\bar{T} = 5$ improvement

Seed $(2,1,4,3)$ $\bar{T} = 5$
Neighbourhood generated:
         $(1,2,4,3)$ $\bar{T} = 25/4$
         $(4,1,2,3)$ $\bar{T} = 7$
         $(3,1,4,2)$ $\bar{T} = 33/4$
         $(2,4,1,3)$ $\bar{T} = 22/4$
         $(2,3,4,1)$ $\bar{T} = 6$
         $(2,1,3,4)$ $\bar{T} = 23/4$

No improvement so select (2, 1, 4, 3) as the heuristically generated schedule. In this case we have found an optimal schedule, but in general optimality is not assured.

10. On Problem 7.6.2:

Palmer's method gives $(3,4,1,2)$ with $C_{max} = 49$ which is
                                                    optimal

Campbell, Dudek and Smith's method
  gives $(3,4,1,2)$ as well
Dannenbring's method gives $(3,4,2,1)$ with $C_{max} = 50$.

On Problem 7.6.3:
Palmer's method gives $(4,3,1,2)$ with $C_{max} = 37$
Campbell, Dudek and Smith's method
  gives $(3,4,2,1)$ with $C_{max} = 38$
Dannenbring's method gives $(4,3,2,1)$ with $C_{max} = 36$ which is
                                                    optimal.

Note that all methods may find non optimal schedules.

11. Suppose J_k obeys condition (i) of Theorem 5.4. Then $\min\{a_k, b_k\} = a_k \leqslant \min\{a_i, b_i\}$ for all J_i. Assume $a_k < b_k$; then $\text{sign}(a_k - b_k) = -1$. It follows that

$$\frac{\text{sign}(a_k - b_k)}{\min\{a_k, b_k\}} \leqslant \frac{\text{sign}(a_i - b_i)}{\min\{a_i, b_i\}} \quad \text{for all} \quad J_i.$$

Hence this sorting algorithm places J_k first in the sequence. If $a_k = b_k$, then J_k also obeys condition (ii) of Theorem 5.4. If J_k obeys condition (ii) (whether or not $a_k = b_k$), then a similar argument shows that the sorting algorithm places J_k last in the processing sequence.

The equivalence of this sorting algorithm and Johnson's algorithm follows immediately.

11.6 PROBLEMS

1. Simply draw the Gantt diagram to obtain this result.

2. The hint is substantial. The difficulty, although only a very minor one, is in showing that

$$\sum_{i=1}^{n} p_i \geq 2(A_k(I) - p^*) + p^*.$$

But one machine finishes at p^* and, because list processing is used, the other must finish between $(A_k(I) - p^*)$ and p^*. Thus, since all jobs are processed we must have

$$\begin{aligned}\sum_{i=1}^{n} p_i &= \text{sum of final completion times on each machine} \\ &\geq A_k(I) + (A_k(I) - p^*).\end{aligned}$$

The result follows.

3. *Step 1* Find the largest j such that $T(n, j) \neq 0$. Set k to $T(n, j)$. Place J_k in S. Set l to $(j - p_k)$.
   Step 2 Find the smallest i such that $T(i, l) \neq 0$. Set k to $T(i, l)$. Place J_k in S. Reset l to $(j - p_k)$.
   Step 3 If $l = 0$, stop. Otherwise, go to *Step 2*

In the problem the following values of $T(i, j)$ are found

		j															
$T(i,j)$		1	2	3	4	5	6	7	8	9	10	11	12	13	14	15	16
	1	0	0	0	0	1	0	0	0	0	0	0	0	0	0	0	0
	2	0	0	0	0	1	0	0	2	0	0	0	0	2	0	0	0
i	3	0	0	0	0	1	0	0	2	3	0	0	0	2	3	0	0
	4	0	4	0	0	1	0	4	2	3	4	4	0	2	3	4	4
	5	0	4	0	0	1	5	4	5	3	4	5	0	5	5	5	5
	6	0	4	6	0	6	5	4	6	6	6	6	6	6	6	5	6

Applying the algorithm to find S, the following is found:
Step 1 $j = 16$; $k = 6$; J_6 placed in S; l set to $(16 - 3) = 13$.
Step 2 $i = 2$; $k = 2$; J_2 placed in S; l set to $(13 - 8) = 5$.
Step 3 $l = 5 > 0$.

Step 2 $i = 1$; $k = 1$; J_1 placed in S; l set to $(5 - 5) = 0$.
Step 3 $l = 0$, stop.

Thus $S = \{J_1, J_2, J_6\}$ are assigned to one machine and the remaining jobs are assigned to the other. C_{max} for the schedule is $(33 - 16) = 17$.

4. For each job let $P_i = \Sigma_{j=1}^m p_{ij}$. Let S be the schedule produced by A. Since S is busy it must complete all the processing by $\Sigma_{i=1}^n \Sigma_{j=1}^m p_{ij} = \Sigma_{i=1}^n P_i$; otherwise there must be some time when all the machines are idle. Thus for all jobs $F_i \leq \Sigma_{i=1}^n P_i$.

Hence
$$\bar{F} \leq \sum_{i=1}^n P_i, \quad \text{i.e. } A(I) \leq \sum_{i=1}^n P_i.$$

Now for the optimal schedule $F_i \geq P_i$, since no job can complete before all its processing has been done.

Hence
$$\text{OPT}(I) \geq \frac{1}{n} \sum_{i=1}^n P_i$$

$$\Rightarrow \frac{A(I)}{\text{OPT}(I)} \leq n \text{ as required.}$$

Noting that (i) a semi-active schedule is busy and (ii) Johnson's algorithm always produces a semi-active schedule, we see from Problem 5.8.5 that this bound may be approached arbitrarily closely. (This performance guarantee is taken from Gonzalez and Sahni, 1978).

5. Assume that the jobs are indexed so that

$$P_1 \leq P_2 \leq P_3 \leq \ldots \leq P_n.$$

Thus the SPT schedule is simply $(J_1, J_2, J_3, \ldots, J_n)$. It follows that under A, $F_i \leq \Sigma_{k=1}^i P_k$ for all jobs J_i.

Hence
$$A(I) \leq \frac{1}{n} \sum_{i=1}^n \left(\sum_{k=1}^i P_k \right)$$

Under an optimal schedule let the jobs complete in the order $(J_{i(1)}, J_{i(2)}, \ldots, J_{i(n)})$. Then, before $J_{i(K)}$ can complete, at least $\Sigma_{k=1}^K P_{i(k)}$ units of processing must be completed on the m machines. Hence under the optimal schedule

$$F_{i(K)} \geq \frac{1}{m} \sum_{k=1}^K P_{i(K)}$$

$$\Rightarrow \quad \text{OPT}(I) \geq \frac{1}{n} \frac{1}{m} \sum_{K=1}^n \left(\sum_{k=1}^K P_{i(K)} \right)$$

Next remember Theorem 3.3 on SPT scheduling for single machine problems. Applying that theorem to this problem as if it were a single machine problem with processing times $P_1, P_2, \ldots, P_n$, we find that the SPT sequence $(J_1, J_2, \ldots, J_n)$ must minimise the double sum $\Sigma_{k=1}^{n} (\Sigma_{k=1}^{K} P_{i(K)})$. Hence

$$OPT(I) - \frac{1}{n} \frac{1}{m} \sum_{k=1}^{n} \left(\sum_{k=1}^{K} P_{i(K)} \right) \geq \frac{1}{n} \frac{1}{m} \sum_{i=1}^{n} \left(\sum_{k=1}^{i} P_k \right) \geq \frac{1}{m} A(I)$$

$$\Rightarrow R_A(I) \leq m \text{ as required.}$$

(This performance guarantee is taken from Gonzalez and Sahni, 1978).

6. Suppose such a polynomial time approximation algorithm exists. Then

$$\frac{A(I)}{OPT(I)} < 1 + \frac{1}{K} = \frac{K+1}{K}$$

Suppose that $A(I) \geq K + 1 \Rightarrow OPT(I) > K$.
Suppose that $A(I) < K + 1 \Rightarrow A(I) \leq K$, since $A(I)$ must be integral

$$OPT(I) \leq K \text{ since } OPT(I) \leq A(I).$$

Hence the question 'Given I, is $OPT(I) \leq K$?' is equivalent to 'Given I, is $A(I) \leq K$'. Since the first question is NP-complete, the equivalent second question cannot be answered in polynomial time. Thus a polynomial time approximation algorithm A such that $R_A(I) < 1 + 1/K$ cannot exist.

References

Adolphson, D. L. (1977) Single machine job sequencing with precedence constraints. *SIAM J. Computing*, **6**, 40–54.

Agin, N. (1966) Optimum seeking with branch and bound. *Mgmt. Sci.*, **13**, B 176–B 185.

Aho, A. V., Hopcroft, J. E. and Ullman, J. D. (1974) *The Design and Analysis of Computer Algorithms*, Addison-Wesley, Reading, Mass.

Akers, S. B. (1956) A graphical approach to production scheduling problems. *Ops. Res.*, **4**, 244–245

Almond, M. (1965) An algorithm for constructing university timetables. *Comput. J.*, **8**, 331–340.

Arumugam, V. and Ramani, S. (1980) Evaluation of value time sequencing rules in a real world job-shop. *J. Opl. Res. Soc.*, **31**, 895–904.

Ashour, S. (1967) A decomposition approach for the machine scheduling problem. *Int. J. Prod. Res.*, **6**, 109–122.

Ashour, S. (1970a) An experimental investigation and comparative evaluation of flow-shop scheduling techniques. *Ops. Res.*, **12**, 541–549.

Ashour, S. (1970b) A branch and bound algorithm for the flow-shop scheduling problem. *A.I.I.E. Trans.*, **2**, 172–176.

Baker, K. R. (1974) *Introduction to Sequencing and Scheduling.* John Wiley, New York.

Baker, K. R. (1975) A comparative survey of flow-shop algorithms *Ops. Res.*, **23**, 62–73.

Baker, K. R., Lawler, E. L., Lenstra, J. K. and Rinnooy Kan, A. H. G. (1980) Pre-emptive scheduling of a single machine to minimise maximum cost subject to release dates and precedence constraints. Mathematisch Centrum, Amsterdam, Preprint No. BW128/80.

Baker, K. R. and Nuttle, H. L. W. (1980). Sequencing independent jobs within a single resource. *Nav. Res. Logist. Q*, **27**, 499–510.

Baker, K. R. and Schrage, L. E. (1978) Finding an optimal sequence by dynamic programming: an extension to precedence-related tasks. *Ops. Res.*, **26**, 111–120.

Bakshi, M. S. and Arora, S. R. (1969). The sequencing problem. *Mgmt. Sci.* **16**, B247–263.

Bansal, S. P. (1979) On lower bounds in permutation flow-shop problems. *Int. J. Prod. Res.*, **17**, 411–418.

Bansal, S. P. (1980) Single machine scheduling to minimise weighted sum of completion times with secondary criterion—a branch and bound approach. *Eur. J. Opl. Res.*, **5**, 177–181.

Bellman, R. (1957) *Dynamic Programming.* Princeton University Press.

Bennett, G. B. and Byrd, J. (1976) A trainable heuristic procedure for the assembly line balancing problem. *A.I.I.E. Trans.*, **8**, 195–201.

Bennington, G. E. and McGinnis, L. F. (1973). A critique of project planning with constrained resources. In Elmaghraby (1973).

Berry, W. L. and Finlay, R. A. (1976) Critical ratio scheduling with queue waiting time information: an experimental analysis. *A.I.I.E. Trans.*, **8**, 161–168.

Blazewicz, J., Lenstra, J. K. and Rinnooy Kan, A. H. G. (1980). Scheduling subject to resource constraints: classification and complexity. Mathematisch Centrum, Amsterdam, Preprint No. BW127/80.

Bowman, E. H. (1959) The schedule—sequencing problem. *Ops. Res.*, **7**, 621–624.

Brooks, G. H. and White, C. R. (1965). An algorithm for finding optimal or near optimal solutions to the production scheduling problem. *J. Ind. Eng.*, **16**, 34–40.

Burkard, R. E. (1980) Remarks on some scheduling problems with algebraic objective functions. *Ops. Res. Verfahren.*, **32**, 63–77.

Burns, R. N. (1976) Scheduling to minimise the weighted sum of completion times with secondary criteria. *Nav. Res. Logist. Q.*, **23**, 125–129.

Campbell, H. G., Dudek, R. A. and Smith, M. L. (1970) A heuristic algorithm for the n-job, m-machine sequencing problem. *Mgmt. Sci.*, **16**, B630–B637.

Coffman, E. G. Jr., Ed. (1976) *Computer and Job-shop Scheduling Theory*. John Wiley, New York.

Conolly, B. (1975) *Lecture Notes on Queueing Systems*. Ellis Horwood, Chichester.

Conolly, B. (1981) *Techniques in Operational Research*. Ellis Horwood, Chichester.

Conway, R. W., Maxwell, W. L. and Miller, L. W. (1967) *Theory of Scheduling*. Addison-Wesley, Reading, Mass.

Cook, S. A. (1971) The complexity of theorem proving procedures. In *Proceedings of the Third Annual ACM Symposium on the Theory of Computing*. Association of Computing Machinery, New York, 151–158.

Corwin, B. D. and Esogbue, A. O. (1974) Two machine flow-shop scheduling problems with sequence dependent set-up times: a dynamic programming approach. *Nav. Res. Logist. Q.*, **21**, 515–524.

Daellenbach, H. and George, J. (1979). *Operational Research Techniques*. Allyn and Bacon.

Dannenbring, D. G. (1977) An evaluation of flow-shop sequencing heuristics. *Mgmt. Sci.*, **23**, 1174–1182.

Dantzig, G. B. (1960). A machine-job scheduling model. *Mgmt. Sci.*, **6**, 191–196.

Dar-el, E. M. (Mansoor) (1975) Solving large scale single-model assembly line balancing problems—a comparative study. *A.I.I.E. Trans.* **7**, 302–310.

Dar-el, E. M. (Mansoor) (1978) Rejoinder to Gehrlein and Patterson (1978) *A.I.I.E. Trans.*, **10**, 112.

Davies, E. W. (1973) Project scheduling under resource constraints—historical review and categorisation of procedures. *A.I.I.E. Trans.*, **5**, 297–313.

Elmaghraby, S. E., Ed. (1973) *Symposium on the Theory of Scheduling and its Applications*. Springer, New York.

Feller, W. (1957) *An Introduction to Probability Theory and its Applications*. Volume 1. John Wiley, New York.

Fisher, M. L. (1980) Worst case analysis of heuristic algorithms. *Mgmt. Sci.*, **26**, 1–17.

Fisher, M. L. and Jaikumar, R. (1978). An algorithm for the space-shuttle scheduling problem. *Ops. Res.*, **26**, 166–182.

Garey, M. R., Graham, R. L. and Johnson, D. S. (1978) Performance guarantees for scheduling algorithms. *Ops. Res.*, **26**, 3–21.

Garey, M. R. and Johnson, D. S. (1978) Strong NP-completeness results: motivation, examples and implications. *J. Assoc. Comput. Mach.*, **25**, 499–508.

Garey, M. R. and Johnson, D. S. (1979) *Computers and Intractability: A Guide to the Theory of NP-Completeness*. Freeman, San Francisco.

Garey, M. R., Johnson, D. S. and Sethi, R. R. (1976) The complexity of flow-shop and job-shop scheduling. *Math. Ops. Res.*, **1**, 117–129.

Garfinkel, R. S. and Nemhauser, G. L. (1972) *Integer Programming*. John Wiley.

Gehrlein, W. V. and Patterson, J. H. (1975) Sequencing for an assembly line with integer task times. *Mgmt Sci.*, **9**, 1064–1070.

Gehrlein, W. V. and Patterson, J. H. (1978) Balancing a single-model assembly line: comments on a paper by E. M. Dar-el (Mansoor) *A.I.I.E. Trans.*, **10**, 109–112.

Gelders, L. F. and Kleindorfer, P. R. (1974) Coordinating aggregate and detailed scheduling in a one-machine job-shop, Part I, Theory *Ops. Res.*, **22**, 46–60.

Gelders, L. F. and Kleindorfer, P. R. (1975) Coordinating aggregate and detailed scheduling in a one-machine job-shop, Part II, Computation and Structure. *Ops. Res.*, **23**, 312–324.

Gelders, L. F. and Sambandam, N. (1978) Four single heuristics for scheduling a flow-shop. *Int. J. Prod. Res.*, **16**, 221–231.

Geoffrion, A. M. (1968) Proper efficiency and the theory of vector maximisation. *J. Math. Anal. Appl.*, **22**, 618–630.

Gere, W. S. (1966) Heuristics in job-shop scheduling. *Mgmt. Sci.*, **13**, 167–190.

Giffler, B. and Thompson, G. L. (1960) Algorithms for solving production scheduling problems. *Ops. Res.*, **8**, 487–503.

Giglio, R. and Wagner, H. (1964) Approximate solution for the three-machine scheduling problem. *Ops. Res.*, **12**, 305–324.

Gittins, J. C. (1979) Bandit processes and dynamic allocation indices. *J. Roy. Statist. Soc.*, **B41**, 148–177.

Gittins, J. C. (1981) Multiserver scheduling of jobs with increasing completion rates *J. Appl. Prob.* **18**, 321–324.

Gittins, J. C. and Nash, P. (1977) Scheduling queues and dynamic allocation indices. *Proc. EMS. Prague* 1974. Czechoslovak Academy of Sciences, Prague 191–202.

Glazebrook, K. D. (1980) On single machine sequencing with order constraints. *Nav. Res. Logist. Q.*, **27**, 123–130.

Glazebrook, K. D. and Nash, P. (1976) On multiserver stochastic scheduling. *J. Roy. Statist. Soc.*, **B38**, 67–72.

Gonzalez, M. J. (1977) Deterministic processor scheduling. *Assoc. Comput. Mach. Surveys.*, **9**, 173–204.

Gonzalez, T. and Sahni, S. (1976) Open-shop scheduling to minimise finish time. *J. Assoc. Comput. Mach.*, **23**, 655–679.

Gonzalez, T. and Sahni, S. (1978) Flow-shop and job-shop schedules: complexity and approximation. *Ops. Res.*, **26**, 36–52.

Graham, R. L. (1969) Bounds on multiprocessing timing anomalies. *SIAM. J. Appl. Math.*, **17**, 416–429.

Graham, R. L., Lawler, E. L., Lenstra, J. K. and Rinnooy Kan, A. H. G. (1979). Optimisation and approximation in deterministic sequencing and scheduling: a survey. *Ann. Discrete. Math.*, **5**, 287–326.

Greenberg, H. H. (1968) A branch and bound solution to the general scheduling problem. *Ops. Res.*, **16**, 353–361.

Gupta, J. N. D. (1971) A functional heuristic algorithm for the flow-shop scheduling problem. *Opl. Res. Q.*, **22**, 39–47.

Gupta, J. N. D. (1972) Heuristic algorithms for multi-stage flow-shop scheduling problems. *A.I.I.E. Trans.*, **4**, 11–19.

Gupta, J. N. D. (1975) A search algorithm for the generalised flow-shop scheduling problem. *Comput. and Ops. Res.*, **2**, 83–90.

Gupta, J. N. D. and Dudek, R. A. (1971) An optimality criterion for flow-shop schedules. *A.I.I.E. Trans.*, **3**, 199–205.

Held, M. and Karp, R. M. (1962) A dynamic programming approach to sequencing problems. *J. SIAM*, **10**, 196–210.

Heller, J. and Logemann, G. (1962) An algorithm for the construction and evaluation of feasible schedules. *Mgmt. Sci.*, **8**, 168–183.

Hindelang, T. J. and Muth, J. F. (1979) A dynamic programming algorithm for decision CPM networks. *Ops. Res.*, **27**, 225–241.

Holloway, C. A. and Nelson, R. T. (1973) Alternative formulation of the job-shop problem with due-dates. *Mgmt. Sci.*, **20**, 65–75.

Holloway, C. A. and Nelson, R. T. (1974a) Job-shop scheduling with due-dates and overtime capability. *Mgmt. Sci.*, **21**, 68–78.

Holloway, C. A. and Nelson, R. T. (1974b) Job-shop scheduling with due-dates and variable processing times. *Mgmt. Sci.*, **20**, 1264–1275.

Holloway, C. A. and Nelson, R. T. (1975) Job-shop scheduling with due dates and operation overlap feasibility. *A.I.I.E. Trans.*, **7**, 16–20.

Horn, W. A. (1974) Some simple scheduling algorithms. *Nav. Res. Logist. Q.*, **21**, 177–185.

Huckert, K., Rhode, R. Roglin, O. and Weber, R. (1980) On the interactive solution of a multi-criteria scheduling problem. *Z. Ops. Res.* **24**, 47–60.

Hurrion, R. D. (1978) An investigation of visual interactive simulation methods using the job-shop scheduling problem. *J. Opl. Res. Soc.* **29**, 1085–1094.

Hwang, C-L. and Masud, A. S. M. (1979) *Multi-Objective Decision Making—Methods and Applications*. Springer Verlag.

Ignall, E. and Schrage, L. E. (1965) Application of the branch and bound technique to some flow-shop problems. *Ops. Res.*, **13**, 400–412.

Jackson, J. R. (1965) Scheduling a production line to minimise maximum tardiness. Research Report, University of California at Los Angeles.

Jaiswal, N. K. *Priority Queues*. Academic Press.

Jeremiah, B., Lalchandani, A. and Schrage, L. (1964) Heuristic rules towards optimal scheduling. Research Report, Department of Industrial Engineering, Cornell University.

Johnson, S. M. (1954) Optimal two- and three-stage production schedules with set-up times included. *Nav. Res. Logist. Q.*, **1**, 61–68.

Kafura, D. G. and Shen, V. Y. (1977) Task scheduling on a multiprocessor system with independent memories. *SIAM J. Comput.* **6**, 167–187.

Kao, E. P. C. (1979) Computational experience with a stochastic assembly line balancing algorithm. *Comput. and Ops. Res.*, **6**, 79–86.

Kao, E. P. C. (1980) A multiple decision theoretic approach to one machine scheduling problems. *Comput. and Ops. Res.*, **7**, 251–259.

Kao, E. P. C. (1981) Erratum and comments on Kao (1979). *Comput. and Ops. Res.* **8**, 58.

Karp, R. M. (1972) Reducibility among combinatorial problems. In *Complexity of Computer Computations*, Miller, R. E. and Thatcher, J. W. Eds. Plenum Press, New York, 95–103.

Karp, R. M. (1975) The fast approximate solution of hard combinatorial problems. In *Proceedings of the Sixth South Eastern Conference on Combinations, Graph Theory and Computing*. Utilitas Mathematica Publishing Co., Winnipeg, 15–34.

References

Kaufmann, A. and Desbazeille, G. (1969) *The Critical Path Model*. Gordon and Breach, New York.

Keeney, R. L. and Raiffa, H. (1976) *Decisions with Multiple Objectives*. John Wiley, New York.

Kelley, J. (1969) Critical path planning and scheduling: mathematical basis. *Ops. Res.*, **9**, 296–320

Khachian, L. G. (1979) A polynomial time algorithm in linear programming. *Dokl. Akad. Nauk. SSSR* **224**, 1093–1096 (In Russian. Translation: Soviet Math. Dokl. **20**, 191–194.)

King, J. R. (1975) *Production Planning and Control*. Pergamon Press, Oxford.

King, J. R. and Spachis, A. S. (1980) Heuristics for flow-shop scheduling. *Int. J. Prod. Res.*, **18**, 345–357.

Kise, H., Ibraki, T. and Mine, H. (1978) A solvable case of the one machine scheduling problem with ready times and due dates. *Ops. Res.* **26**, 121–126.

Kise, H., Ibaraki, T. and Mine, H. (1979) Performance analysis of six approximation algorithms for the one machine maximum lateness scheduling problem with ready times. *J. Ops. Res. Soc. of Japan*, **22**, 205–223.

Kise, H., Uno, M. and Kabata, M. (1979) Approximation algorithms for identical parallel machine scheduling problems with ready times and precedence constraints. *Memoirs of Faculty of Industrial Arts, Kyoto Technical University, Science and Technology*, **28**, 64–72.

Klee, V. (1980) Combinatorial Optimisation: What is the state of the art? *Math. Ops. Res.*, **5**, 1–26.

Kohler, W. H. and Steiglitz, K. (1976) Enumerative and iterative computational approaches. In Coffman, E. G. Ed. (1976). 229–287.

Labetoulle, J. and Lawler, E. L. (1978) On pre-emptive scheduling of unrelated parallel processors by linear programming. *J. Assoc. Comput. Mach.*, **25**, 612–619.

Lageweg, B. J., Lenstra, J. K. and Rinnooy Kan, A. H. G. (1977). Job-shop scheduling in implicit enumeration. *Mgmt. Sci.*, **24**, 441–450.

Lageweg, B. J., Lenstra, J. K. and Rinnooy Kan, A. H. G. (1978). A general bounding scheme for the permutation flow-shop problem. *Ops. Res.*, **26**, 53–67.

Lawler, E. L. (1964) On scheduling problems with deferral costs. *Mgmt. Sci.*, **11**, 280–288.

Lawler, E. L. (1973) Optimal sequencing of a single machine subject to precedence constraints. *Mgmt. Sci.*, **19**, 544–546.

Lawler, E. L. (1978) Sequencing jobs to minimise total weighted completion time subject to precedence constraints. *Ann. Discrete. Math. 2*, 75–90.

Lawler, E. L. (1979) Efficient implementation of dynamic programming algorithms for sequencing problems. Preprint BW106/79 Mathematisch Centrum, Amsterdam.

Lawler, E. L. and Moore, J. M. (1969) A functional equation and its application to resource allocation and sequencing problems *Mgmt. Sci.*, **16**, 77–84.

Lawler, E. L. and Wood, D. E. (1966) Branch and bound methods: a survey. *Ops. Res.*, **14**, 1098–1112.

Lawrie, N. L. (1969) An integer linear programming model of a school management problem. *Comput. J.*, **12**, 307–316.

Lenstra, J. K. (1977) *Sequencing by Enumerative Methods*. Mathematisch Centrum, Amsterdam.

Lenstra, J. K. and Rinnooy, Kan, A. H. G. (1978) Complexity of scheduling under precedence constraints. *Ops. Res.*, **26**, 22–35.

Lenstra, J. K. and Rinnooy Kan, A. H. G. (1979) Computational complexity of discrete optimisation problems. *Ann. Discrete. Math.*, **4**, 121–140.

Lenstra, J. K. and Rinnooy Kan, A. H. G. (1980) An introduction to multiprocessor scheduling. Mathematisch Centrum, Amsterdam. Preprint BW121/80.

Lenstra, J. K., Rinnooy Kan, A. H. G. and Brucker, P. (1977) Complexity of machine scheduling problems. *Ann. Discrete. Math.*, **1**, 343–362.

Lockett, A. G. and Muhlemann, A. P. (1972) A scheduling problem involving sequence-dependent changeover times. *Ops. Res.*, **20**, 895–902.

Lockyer, K. G. (1969) *An Introduction to Critical Path Analysis*. Pitman Publishing Co., London.

Lomnicki, Z. (1965) A branch and bound algorithm for the exact solution of the three machine scheduling problem. *Ops. Res. Q.* **16**, 89–100.

McMahon, G. B. and Burton, P. G. (1967) Flow-shop scheduling with the branch and bound method. *Ops. Res.*, **15**, 473–481.

Manne, A. S. (1960) On the job-shop scheduling problem. *Ops. Res.*, **8**, 219–223.

Mellor, P. (1966) A review of job-shop scheduling. *Opl. Res. Q.*, **17**, 161–171.

Monma, C. L. (1979) The two machine maximum flow time problem with series-parallel precedence constraints: an algorithm and extensions. *Ops. Res.*, **27**, 792–798.

Monma, C. L. (1980) Sequencing to minimise the maximum job-cost. *Ops. Res.*, **28**, 942–951.

Moore, J. M. (1968) An n-job, one machine sequencing algorithm for minimising the number of late jobs. *Mgmt. Sci.*, **15**, 102–109.

Moore, L. J., Taylor, B. W., Clayton, E. R. and Lee, S. M. (1978) Analysis of a multicriteria project crashing model. *A.I.I.E. Trans.* **10**, 163–169.

Moran, P. A. P. (1968). *An Introduction to Probability Theory*, Oxford University Press, Oxford.

Muth, J. F. (1963) The effect of uncertainty in job times on optimal schedules. In Muth and Thompson (1963), 300–307.

Muth, J. F. and Thompson, G. L. (Eds.) (1963) *Industrial Scheduling*. Prentice Hall, Englewood Cliffs, New Jersey.

Nelson, R. T., Holloway, C. A. and Wong, R. M.-L. (1977). Centralised scheduling and priority implementation heuristics for a dynamic job-shop model. *A.I.I.E. Trans.*, **9**, 95–102.

Page, E. S. (1961) An approach to the scheduling of jobs on machines. *J. Roy. Statist. Soc.*, **B23**, 484–492.

Palmer, D. S. (1965) Sequencing jobs through a multi-stage process in the minimum total time—a quick method of obtaining a near optimum. *Opl. Res. Q.*, **16**, 101–107.

Panwalker, S. S. and Iskander, W. (1977) A survey of scheduling rules. *Ops. Res.*, **25**, 45–61.

Panwalker, S. S. and Khan, A. W. (1976) An ordered flow-shop sequencing problem with mean completion time criterion. *Int. J. Prod. Res.*, **14**, 631–635.

Panwalker, S. S. and Woollam, C. R. (1979) Flow-shop scheduling problem with no in-process waiting: a special case. *J. Opl. Res. Soc.* **30**, 661–664.

Panwalker, S. S. and Woollam, C. R. (1980) Ordered flow-shop problems with no in-process waiting: further results. *J. Opl. Res. Soc.*, **31**, 1039–1043.

Panwalker, S. S., Dudek, R. A. and Smith, M. L. (1973) Sequencing research and the industrial problem. In Elmaghraby (1973).

Papadimitriou, C. H. and Kanellakis, P. C. (1980) Flow-shop scheduling with limited temporary storage. *J. Assoc. Comput. Mach.*, **27**, 533–549.

Papoulias, D. B. (1980) The assignment-to-days problem of a school timetable: a heuristic approach. *Eur. J. Opl. Res.*, **4**, 31–41.

Papoulias, D. B. and Panayiotopoulos, J-C. (1979). Considerations on the requirements matrix of a non-regular school timetabling problem. *Eur. J. Opl. Res.*, **3**, 382–385.

Potts, C. N. (1978) An adaptive branching rule for the permutation flow-shop problem. Report No. BW91/78. Mathematisch Centrum, Amsterdam.

Pritsker, A. A. B., Miller, L. W. and Zinkl, R. J. (1971) Sequencing n products involving m independent jobs on m machines. *A.I.I.E. Trans.* **3**, 49–60.

Randolph, B. H., Swinson, G. and Ellingsen, C. (1973). Stopping rules for sequencing problems. *Ops. Res.*, **21**, 1309–1315.

Rinnooy Kan, A. H. G. (1976) *Machine Scheduling Problems: Classification, Complexity and Computations.* Martinus Nijhoff, The Hague, Holland.

Rinnooy Kan, A. H. G., Lageweg, B. J. and Lenstra, J. K. (1975) Minimising total costs in one machine scheduling. *Ops. Res.*, **23**, 908–927.

Sahney, V. K. (1972) Single server, two machine sequencing with switching time *Ops. Res.*, **20**, 24–36.

Sahni, S. (1976) Algorithms for scheduling independent tasks. *J. Assoc. Comput. Mach.*, **23**, 116–127.

Sahni, S. (1977) General techniques for combinatorial optimisation *Ops. Res.* **25**, 920–936.

Sahni, S. and Cho, Y. (1979) Complexity of scheduling shops with no-wait in process. *Math. Ops. Res.*, **4**, 448–457.

Salvador, M. S. (1973) A solution to a special class of flow-shop scheduling problems. In Elmaghraby (1973) 83–91.

Schrage, L. (1970) Solving resource-constrained network problems by implicit enumeration—non pre-emptive case. *Ops. Res.* **18**, 263–278.

Schrage, L. (1972) Solving resource-constrained network problems by implicit enumeration—pre-emptive case. *Ops. Res.*, **20**, 668–677.

Schrage, L. and Baker, K. R. (1978) Dynamic programming solution of sequencing problems with precedence constraints. *Ops. Res.*, **26**, 444–449.

Shapiro, R. D. (1980) Scheduling coupled tasks, *Nav. Res. Logist. Q.*, **27**, 489–497.

Sidney, J. B. (1973) An extension of Moore's due date algorithm. In Elmagehraby, S. R. (1973) 393–398.

Sidney, J. B. (1975) Decomposition algorithms for single machine sequencing with precedence relations and deferral costs. *Ops. Res.*, **23**, 283–298.

Sidney, J. B. (1977) Optimal single machine scheduling with earliness and tardiness penalties. *Ops. Res.*, **25**, 62–69.

Sidney, J. B. (1979) The two machine maximum flow-time problem with series-parallel precedence constraints. *Ops. Res.*, **27**, 782–791.

Silver, E. A., Vidal, R. V. and De Werra, D. (1980). A tutorial on heuristic methods. *Eur. J. Opl. Res.*, **5**, 153–162.

Smeds, P. A. (1980) Methods for checking the consistency of precedence constraints. *A.I.I.E. Trans.*, **12**, 171–178.

Smith, M. L., Panwalker, S. S. and Dudek, R. A. (1976) Flow-shop sequencing problem with ordered processing time matrices: a general case. *Nav. Res. Logist. Q.*, **22**, 481–486.

Smith, W. E. (1956) Various optimizers for single state production. *Nav. Res. Logist. Q.*, **3**, 59–66.

Spachis, A. S. and King, J. R. (1979) Job-shop scheduling with local neighbourhood search. *Int. J. Prod. Res.*, **17**, 507–526.

Sphicas, G. P. and Silverman, F. N. (1976) Deterministic equivalents for stochastic assembly line balancing. *A.I.I.E. Trans.*, **8**, 280–282.

Stinson, J. P., Davis, E. W. and Khumawala, B. M. (1978) Multiple resource-constrained scheduling using branch and bound. *A.I.I.E. Trans*, **10**, 252–259.

Story, A. E. and Wagner, H. M. (1963) Computational experience with integer programming for job-shop scheduling. In Muth and Thompson, Eds. (1963) 207–219.

Sturm, L. B. J. M. (1970) A simple optimality proof of Moore's sequencing algorithm. *Mgmt. Sci.*, **17**, B116–B118.

Szwarc, W. (1973) Optimal elimination methods in the m × n flow-shop scheduling problem. *Ops. Res.*, **21**, 1250–1259.

Szwarc, W. (1977) Optimal two machine orderings in the 3 × n flow-shop problem. *Ops. Res.* **25**, 70–77.

Ullman, J. D. (1976). Complexity of sequencing problems. In Coffman, E. G. (Ed.) (1976) 139–164.

Van Assche, F. and Herroelen, W. S. (1978) An optimal procedure for the single-model deterministic assembly line balancing problem. *Eur. J. Opl. Res.*, **3**, 142–149.

Van Wassenhove, L. N. and Baker, K. R. (1980) A bicriterion approach to time/cost trade-offs in sequencing. Paper presented at the 4th European Congress on Operational Research, Cambridge, England, July 22–25, 1980. Submitted to A.I.I.E. Trans.

Van Wassenhove, L. N. and Gelders, L. F. (1978) Four solution techniques for a general one-machine scheduling problem: a comparative study. *Eur. J. Opl. Res.*, **2**, 281–290.

Van Wassenhove, L. N. and Gelders, L. F. (1980) Solving a bicriterion scheduling problem. *Eur. J. Opl. Res.*, **4**, 42–48.

Vickson, R. G. (1980) Choosing the job sequence and processing times to minimise total processing plus flow cost on a single machine. *Ops. Res.* **28**, 1155–1167.

Wagner, H. M. (1959) An integer programming model for machine scheduling. *Nav. Res. Logist. Q.*, **6**, 131–140.

Weide, B. (1977) A survey of analysis techniques for discrete algorithms. *Assoc. Comput. Mach. Surveys.*, **9**, 291–313.

White, C. H. and Wilson, R. C. (1977) Sequence-dependent set-up times and job sequencing. *Int. J. Prod. Res.*, **15**, 191–202.

White, D. J. (1969) *Dynamic Programming.* Oliver and Boyd, Edinburgh.

White, D. J. (1975) A note on faculty timetabling. *Opl. Res. Q.*, **26**, 875–878.

Whittle, P. (1980) Multiarmed bandits and the Gittin's index. *J. Roy. Statist. Soc.*, **B42**, 143–149.

Wilkinson, L. J. and Irwin, J. D. (1971) An improved algorithm for scheduling independent tasks. *A.I.I.E. Trans.* **3**, 239–245.

Wismer, D. A. (1972) Solution of flow-shop scheduling problem with no intermediate queues. *Ops. Res.*, **20**, 689–697.

Yoshida, T. and Hitomi, K. (1979) Optimal two stage production scheduling with set-up times separated. *A.I.I.E. Trans.*, **11**, 261–264.

Yueh Ming-I (1976) On the n job, m machine sequencing problem of a flow-shop. In Operational Research 1975, Haley, K. B. (Ed.), North Holland, Amsterdam.

Symbol Index

Notation which is local to one or perhaps two sections is not included below

Symbol	Page Introduced	Description and Notes
a_i	10	allowance; and
	69	processing time of J_i on M_1.
$A(I)$	177	the value of the performance measure achieved by applying the approximation algorithm A to instance I.
b_i	69	processing time of J_i on M_2.
c_i	110	processing time of J_i on M_3.
C_i	10	completion time of J_i.
d_i	10	due date of J_i.
E_i	10	earliness of J_i.
F_i	10	flow time of J_i.
I_j	11	idle time on M_j.
J_i	5	the job J_i.
$lb(A)$	111	lower bound function.
L_i	10	lateness of J_i.
m	5	the number of machines.
M_j	5	the machine M_j.
n	5	the number of jobs.
n_T	13	the number of tardy jobs.
NP	145	the class NP.
$N_c(t)$	12	the number of jobs completed by time t.
$N_p(t)$	12	the number of jobs actually being processed by time t.
$N_u(t)$	12	the number of jobs still to be completed by time t.
$N_w(t)$	12	the number of jobs waiting at time t.
o_{ij}	6	the operation of processing J_i on M_j.
o_k	158	an indexing of the operations when their association with jobs and machines is of secondary importance.
OPT(I)	177	the value of the performance measure achieved by the optimal schedule for instance I.
p_{ij}	6	processing time of J_i on M_j. (N.B. this notation is used differently in section 4.2).
p_i	35	processing time of J_i in a single machine problem.
P	145	the class P.
r_i	6	ready time of J_i.
$R_A(I)$	177	relative performance of algorithm A on instance I.

Symbol	Page Introduced	Description and Notes
s_j	200	speed factor of M_j.
T_i	10	tardiness of J_i.
W_{ik}	10	waiting time of J_i preceding its k-th operation. (N.B. this notation is modified in Chapter 8: see page 133.).
W_i	10	total waiting time of J_i.
α_i	110	starting time of J_i on M_1.
β_i	110	starting time of J_i on M_2.
γ_i	110	starting time of J_i on M_3.
C_{max}, T_{max}, etc.	11	maximum of quantity.
$\bar{C}$, $\bar{T}$, etc.	11	mean of quantity.
$n/m/A/B$	14	classification of problem.
$\{J_{i(1)}, J_{i(2)}, \ldots, J_{i(K)}\}$	88	an unordered set of jobs.
$(J_{i(1)}, J_{i(2)}, \ldots, J_{i(K)})$	88	a sequence of jobs.
$Q-\{J\}$	88	set of jobs obtained by deleting J from Q.
$f(v)$ is $O(g(v))$	140	f and g are of the same order.
$\Pi_1 \propto \Pi_2$	147	problem Π_1 is polynomially reducible to problem Π_2.

Author Index

Subject Index